Praise for *Matter and Desire*

"The most powerful antidote to our pernicious culture of excessive material consumption is the creation and nurturing of communities, finding happiness in human relationships rather than seeking it in material possessions. At the very heart of community, at all levels of life, we find a fundamental impulse to establish connections. The author of this beautifully written book identifies this yearning for connections with the essence of love. In his philosophical meditations, Andreas Weber deepens the recent scientific advances toward a new systemic understanding of life by investing them with a vital emotional dimension. While the experience of being fully alive is, for him, an erotic experience, it has also been recognized as the very essence of spirituality. An important and inspiring book!"

—FRITJOF CAPRA, author of *The Web of Life*;
coauthor of *The Systems View of Life*

"Andreas Weber is an indispensable voice in ecological and philosophical thought. With fearless probity and autobiographical intimacy, *Matter and Desire* composes the symphonic grand design of desire, relationships, the metaphysics of the body—and much more—as page by page we experience Weber's elegant subversion of all convenient ways of looking at the natural world. This is a timeless yet urgent, and splendid book."

—HOWARD NORMAN, author of
I Hate to Leave This Beautiful Place

"Andreas Weber offers us the best medicine I know for a culture benumbed by dead-end pursuits. Pulsing with life, his work delivers us from the centuries-long dichotomies between mind and matter that have robbed us of vitality, joy, and true purpose. It brings us home to the fertile reciprocities that link us with all forms and levels of life; in so doing, it reflects and reinforces great spiritual teachings of our planet."

—JOANNA MACY, author of *Coming Back to Life*

"A slow tidal wave of change is gathering force and will take us beyond the mechanistic world of Newton toward one of becoming. Andreas Weber's *Matter and Desire* is a passionate evocation of intermingled life surging. He writes with the poetry, care, and insight that urges us forward."

—Stuart Kauffman, professor emeritus, biochemistry and biophysics, University of Pennsylvania; and MacArthur Fellow

"With a dazzling blend of biological rigor and poetic grace, Andreas Weber explains the principles of erotic connection that lie at the heart of life on Earth. It is a journey that transcends the reductionist taxonomies of modern science and explains the transformational role of desire, interdependence, and meaning in the glorious unfolding of natural ecosystems—and in our own lives. Be prepared for a bracing adventure!"

—David Bollier, author of *Think Like a Commoner*

"When Andreas Weber looks on a meadow, he sees 'part of our body, folded outward, ready to be strolled through.' The ocean's tides are 'the way the Earth perceives the moon,' and gravitation is 'the Earth's tender longing for us.' With such graceful, lucid lines Weber invites us to see a world filled with delight and one that yearns, as we do, for contact: the erotics of encounter. Part scientific reflection, part philosophical reverie, part lyrical benediction for the stones and swifts and plants and water ouzels of his beloved Ligurian countryside, *Matter and Desire* is a deeply felt book from a profoundly humane writer."

—Fred Bahnson, author of *Soil and Sacrament*; director, Food, Health, and Ecological Well-Being Program, Wake Forest University School of Divinity

"Every page of Weber's deeply illuminating new book is a passionate journey into the experience of being alive and in relationship. As an emergent 'erotic ecology,' this book is urgently needed medicine for a planet suffering from a shortage of love."

—David Lukas, author of *Language Making Nature*

"Two hundred years ago, John Keats complained that modern science would 'unweave a rainbow.' This visionary and poetic discourse by Andreas Weber achieves the near-miraculous task of reweaving the stunning beauty of the natural world back into the realm of science. Transcending conventional barriers between categories of Western thought, with a style reminiscent of Annie Dillard's *Pilgrim at Tinker Creek*, Weber explores some profound implications of modern biology and physics, presenting his vision of biology as the erotic science with the recognition that to truly experience love, we need to be fully connected to the creativity of life."

—JEREMY LENT, author of *The Patterning Instinct*

"To read this marvellous book is to enter a secret garden where you'll discover a natural world far more alive, sentient, and meaningful than science has so far dared suppose. With luminous prose Weber's 'erotic ecology' charts a path into a new scientific understanding in which atoms, organisms, and entire ecosystems overflow with purpose, interiority, and psyche, lighting up your life, helping you experience reality with freshness and depth of vision. A masterpiece."

—DR. STEPHAN HARDING, author of *Animate Earth*

"A stunning piece of writing, as existential as it is experiential, *Matter and Desire* delves into the 'science of the heart' in compelling prose that frequently dances on the edge of poetry. The book provides vivid depictions of a big love: a near-mystical practice of discovering who we are through the creative energies that surround us and dwell within us. Andreas Weber ably guides his readers on this relational journey, articulating ecological intuitions that may have gone unnoticed yet were always on the tips of our tongues. From the forces of desire within molecules to the mistle thrush's song vibrating in the evening air, Weber offers a bold and convincing case for the physicality of feeling and the 'biology of love.' The result is a profound meditation that bravely explores the subjectivity of a living biosphere and our particular relations within it. If philosophy literally means the love of wisdom, then in *Matter and Desire*, Weber presents the wisdom of love, a reflective account of his intentional free-fall into the embrace of matter."

—GAVIN VAN HORN, director, Cultures of Conservation,
Center for Humans and Nature

MATTER
& DESIRE

An Erotic Ecology

ANDREAS WEBER

Translated by Rory Bradley
Foreword by John Elder

Chelsea Green Publishing
White River Junction, Vermont

Cover photograph, *Smoke and Mirrors 2, 2010*, by Ellie Davies (www.elliedavies.co.uk).

Project Manager: Angela Boyle
Project Editor: Brianne Goodspeed
Copy Editor: Deborah Heimann
Proofreader: Eileen M. Clawson
Indexer: Linda Hallinger
Designer: Melissa Jacobson

Printed in the United States of America.
First printing July 2017.
10 9 8 7 6 5 4 3 2   20 21 22 23

Library of Congress Cataloging-in-Publication Data
Names: Weber, Andreas, 1967– author.
Title: Matter and desire : an erotic ecology / Andreas Weber ; translated by Rory Bradley.
Other titles: Lebendigkeit. English
Description: White River Junction, VT : Chelsea Green Publishing, [2017] | Translated from German.
Identifiers: LCCN 2017005009| ISBN 9781603586979 (paperback) | ISBN 9781603586986 (ebook)
Subjects: LCSH: Human ecology—Philosophy. | Philosophy of nature. | Mind and body. | BISAC: NATURE
    / Ecology. | SCIENCE / Philosophy & Social Aspects. | SCIENCE / Life Sciences / Ecology. |
    PHILOSOPHY / Mind & Body.
Classification: LCC GF21 .W4313 2017 | DDC 304..201.—dc23
LC record available at https://lccn.loc.gov/2017005009

Chelsea Green Publishing
85 North Main Street, Suite 120
White River Junction, VT 05001
(802) 295-6300
www.chelseagreen.com

Nature is on the inside.

PAUL CÉZANNE

Two forces rule the universe: light and gravity.

SIMONE WEIL

*To know oneself means to be oneself, to be master of oneself,
to distinguish oneself, to free oneself from a state of chaos,
to exist as an element of order but of one's own order and
one's own discipline in striving for an ideal. And we
cannot be successful in this unless we also know others,
their history, the successive efforts they have made to be
what they are [...] And we must learn all this without
losing sight of the ultimate aim: to know oneself better
through others and to know others better through oneself.*

ANTONIO GRAMSCI[1]

— *contents* —

— *foreword* —

I n *Matter and Desire: An Erotic Ecology*, Andreas Weber has written a beautiful, timely book. Beautiful, because of his lovingly observed descriptions of swifts and toads, of rivers splashing through rocky landscapes and a pond cradled in a forest. Timely, because he focuses throughout on the ways in which sensory contact with our fellow creatures, as well as with air and water, light and gravity, can deepen our capacity to identify with all of life. Such an experience of passionate affiliation is essential if we in the affluent West and the Pacific Rim are to rein in the consumerist greed that (as Bill McKibben discusses in *Enough*) underlies climate change, habitat destruction, the eradication of species, and the victimization of impoverished human communities around the globe. Politics and technology of course have important roles to play in responding to our present environmental catastrophe. But more fundamental are the values that *motivate* individuals and societies to adopt more just, generous, and compassionate practices, and in doing so to transcend a narrow understanding of self-interest. As Weber illustrates and discusses so compellingly, such motivation can germinate within the "web of mutual transformations and deeply meaningful encounters that are always embodied."

Though Weber calls his book a sequence of "love stories," he scrupulously relates both love and the erotic to ecology, which he calls the "science of the heart." As a biologist who worked closely with the philosopher and neuroscientist Francisco Varela in Paris, he is knowledgeable about the continuity between our bodies and those of all other living beings. But a human experience of desire and satisfaction is most significant, he further argues, as a vivid contact with the outside world that simultaneously amplifies one's sense of inwardness and self-awareness. In this regard, Weber describes love as "the principle of a fulfilling equilibrium between the individual and the whole." This is his way of conveying how certain vibrant experiences of reciprocity allow us to feel both our own life and that of the world in a new, delightfully intense way. (It's worth noting in this regard that the book's original title in German was *Lebendigkeit*, or *Aliveness*.) As powerful as such an equilibrium might be, though, its *ecological* dimension enforces the fact that this principle is only fully manifested in a physical universe of constant change, loss, and death. Such an emphasis both keeps his celebration of love from becoming sentimental or abstract and harmonizes his personal and philosophical overview with the theory of natural selection and other key evolutionary concepts.

"Erotic Ecology," in the subtitle, bears down on this starker, less individual dimension of human desire. As Weber states, Eros was considered a terrifying god by the Greeks—one capable at any moment of driving the wise mad and bringing the powerful low. Though love is often exhilarating, it is also inextricable from our mortality or, to put it another way, from the world's inevitable fatality for every organism. There's no road through life that doesn't lead to death. A great strength of this book is the author's discerning way of incorporating poetry expressive of this tragic view of love. Dylan Thomas, though not cited here, memorably sums up this perception of an overwhelming, erotic current running through and beyond our individual lives:

> *The force that through the green fuse drives the flower*
> *Drives my green age; that blasts the roots of trees*
> *Is my destroyer.*[1]

In addition to his accounts of enlivening moments of contact with landscapes, human beings, and other creatures, and to his discussion of certain visionary scientists and philosophers of science, Andreas Weber also draws on two other relevant lineages. One is the tradition of American nature writing. John Muir is a writer who fascinates him, in part because Muir's emotional, exuberant tone so closely anticipates Weber's own experiences of ecstatic "contact." Weber especially appreciates Muir's hymn, in *The Mountains of California*, to the water ouzel as the genius of place initiating him into the full vitality of a High Sierra stream. A writer of our own day with whom Weber feels a similar rapport is David Abram, author of *The Spell of the Sensuous*. Like Abram, he weaves together personal narrative and lyrical description with dense discussions of ecology and psychology. Whereas Abram's emphasis in his more analytical sections is on Husserl, Merleau-Ponty, and other leading phenomenologists, however, Weber seems to have been particularly influenced by the school of humanistic psychology, including Erich Fromm, Rollo May, and Abraham Maslow. Maslow's concept of "self-actualization" is highly pertinent to Weber's own experience of gaining an expansive awareness of the web of life through experiences of self-transcending love. Though organic life is fragile, and death is inextricable from life, there are nonetheless moments of connection that lift us up into a larger perspective, strengthening our potential for compassion and love.

This is the hopeful heart of *Matter and Desire*. By affirming the realm of arising and submerging forms of which we are a part, we can gain an enlivening, and ultimately a liberating, view of life on Earth. Weber describes the project of "modernity" as having been a five-hundred-year-long attempt "to hide or abolish death," and in that way a removal of human beings from what Forster has called "the pattern that connects." But his book also exemplifies another historical impulse: the Romantic Revolution that began about two and a half centuries ago in Germany and England. In opposition to the elitism, rationalism, positivism, and hierarchical thinking of the Enlightenment, writers like the Brothers Grimm and William Wordsworth asserted the values of physical experience, emotion, organic form, the insights of rural people, and "old wives' tales." This counter-impulse has continued to play out in our own day, through the revolutionary agendas of the anticolonial, civil rights, environmental, women's, and gay liberation movements.

Recognizing the specific continuity between early Romantic "prophets of nature" and conservationists like John Muir, Rachel Carson, and Aldo Leopold may help us move past the emphasis on purity and separation that has sometimes limited the environmental movement's appeal. Precisely because the ecological challenges are now so urgent, especially in relation to climate change, we require a more invitational, inclusive, and personally engaged approach to environmental activism. The context, at such a moment of necessary reorientation, remains a sobering one. There seems little likelihood of quick or easy corrections to our society's current, destructive practices. Even less escapable, of course, are the losses ingrained in all of organic life. Hence, Weber's encapsulation of his vision, late in the book, as "metaphysics in the mood of loss." But this reality lends even greater preciousness to the joy and laughter erupting at moments of reawakening to our kinship with all of life. Through celebrating such deep physical and emotional bonds we prepare ourselves to reclaim our place in nature and to assume the responsibilities that follow from it.

JOHN ELDER

— *preliminary remarks* —

*W*ithout attachments, no life. From cell division to child rearing, we can understand all processes in the biosphere as processes of relationship—and we can learn from them. In these processes, two different positions must be brought into balance such that something altogether new emerges, something that both contains and completely redefines everything that preceded it. This connection of two (or more) different positions in a common cause—one that remains full of contradictions —is perhaps the most general definition of an ecosystem. It is also the precise description of a loving attachment.

This book therefore pursues an ambitious goal: It investigates the principles of reality that we can experience and of which we are a part. But it tries to do so through a science of the heart and not by means of a mere biological description of bodies and their senses. The impetus for this risky undertaking is the conviction that we are currently neglecting reality because our efforts to describe and understand the world are directed away from the experience of being alive and being in relationship. In other words: We consider the practice of love a private matter, rather than an instrument of knowledge.

On the following pages, I describe this reality as the creative, poetic nexus of unfolding freedom toward both individuation and attachments. Traditionally, this drive toward both the self and the fullness of connections answered to the name "Eros." Throughout natural history, reality has unfolded in the form of living systems, in the form of self-organizing molecules, cells, bodies, biotopes, and landscapes; in each of these, the drive, desire, and longing for attachment *and* autonomy is foundational: essential in order to perceive, to continue, and to unfold.

For all these reasons I call my writing in this book an "erotic ecology." Being in the world is primarily an erotic encounter, an encounter of meaning through contact, an encounter of being oneself through the significance of others—humans, lovers, children, but also other beings, companions and competitors. From birth, and probably even before it, we experience the fundamental erotics of being touched by the world, and of touching it in return, as a life-bestowing power. We experience living exchange as fundamental reality. We long to connect with an other—be it word, skin, food, or air—in order to become ourselves. In this experience, we are not separated from the world, but deeply incorporated into it: feeling parts of the whole, which can thus become transparent to itself in a meaningful way. It is precisely this reality, in all of its creative growth, that we wish to preserve—an expressive, meaningful reality of which we are a part.

As we are a part of it, we cannot detach ourselves from it in order to paint an objective picture. But we can express what it means to be the participant in a web of mutual transformations and deeply meaningful encounters that are always embodied. We know what it means to be enmeshed in an erotic partaking in reality, and we can express from the inside what it feels like to be alive. And from this vantage point we might be able to give back to the world what it has most painfully been lacking—the experience of aliveness, and the knowledge that reality is not only an efficient organization of matter, but that it also calls forth interiorities full of meaning and expression. Reality is alive, and it is about being on the inside—in the felt experience of pain and joy. Writing about being alive as an "erotic ecology" means becoming a partisan of poetics and striving for the reality of our enlivened experiences through the connections and the transformations they entail. Therefore, this book describes ecological reality as a relational system. And conversely, it comprehends love as an ecological process.

My conviction is that being alive in an empathetic way is always a practice of love. And only by relearning to understand our existence as a practice of love will we grasp anew the overwhelming ecological and human dilemmas that we face in the middle of the second decade of the twenty-first century and find the means to deal with them differently than we have thus far. Life is the constant, creative transition from controlled situations to new openings that cannot be controlled. Being in tune with life lies somewhere between following rigid principles and improvising on them vividly. Cultivating a practice of love that tries to remain close to the ecological Eros therefore means caring for oneself but also remaining vulnerable, a balanced center always open to new connections. From an ecological perspective, love is a practice of balancing interests that lead to a state of greater aliveness while also accepting failure in advance. A successful attachment always has two sides: living without fear, and learning to die courageously.

Love is an answer to the lack that lies at the heart of aliveness, but it does not compensate for that lack—it transforms it. Love transforms that lack into an excess that produces new contradictions; it is the luminous chasm and the ephemeral mass, freedom in impossibility, the always insufficient answer to the paradox of life: "vivacidad pura" (Octavio Paz)—pure aliveness, experienced from inside the world.

Accordingly, I will tell a series of love stories on the following pages. I will describe and analyze erotic affairs with stones, plants, rivers, animals, people, and words. Through these stories, I will understand the overpowering extent to which reality is determined by the erotic—by the longing for a practice of being meaningfully moved in our embodied existences. I would like to probe the extent to which we have forgotten this reality, and to discover how we might reclaim it.

— *prelude* —

THE CARRYING CAPACITY
OF AIR

And so, no one has more spirit than he has love.

THEODOR LESSING[1]

"There's a rasping sound in my chimney again," my friend says in agitation. Then she has to laugh at herself. She rocks from one leg to the other. She hasn't changed her clothes. In her housecoat, she stands in front of my table outside on the terrace of Walter's Bar.

Balmy air fills the square in the center of the little Italian town in the mountains above the Riviera. The sun has already disappeared, but like its afterimage, a silvery light settles between the houses, as though the night wants to remain illuminated. "Something is clawing and scraping behind the wall again. Could you take another look?" she asks. She has to laugh again. "Maybe a cat fell in there," I say. Motioning to the host that I will pay later, I get up and follow my friend down the warm alleyway to her apartment.

She works at the school. She has a position that does not exist any-more in my native country Germany: She is the caretaker. Her duties include those of the secretary, the housekeeper, and the cleaning woman. And she has to ring the bell punctually for break. But the truth is that she is the school nurse. Time and again, you see students at her table at the end of the corridor. Girls and boys sit there, heads between their arms, oppressed by the burden of learning, by the torment of being a child. Even in the middle of the lesson.

The children are not waiting here for the headmaster, shame-faced because they have done something wrong. Here, they sit at the unofficial school nurse's table whenever they are unhappy, their heads buried in their arms. The caretaker consoles them. Or actually, she doesn't console at all. She laughs. She laughs about their pain and their suffering. The schoolchildren drag themselves to her, crying, and she laughs. And that, precisely, is the medicine. The caretaker laughs, but she isn't laughing at the children; she is laughing at the pain. She laughs about the millionth little unhappiness so warmheartedly and so kindly that it is contagious, and the rage or the pain is lessened.

In her apartment, I climb onto a wobbly chair. She doesn't have more than three, all of them somehow defective. When people come over for a meal, they have to gather the sofa cushions around the table. I pull the balled-up towel out of the hole in the wall intended for a stovepipe. Ashes fall, rustling, onto the floorboards and table. We have to look at each other and laugh again. I already did this once, in the morning—pulled out the towel and stood on a chair on my tiptoes, shining a flashlight into the hole above my head. No luck.

This time I feel around with my hand. Again we hear the scratching sound—louder this time, frantic. I reach deeper into the chimney and touch something smooth, soft, something round and moving. I shiver for a moment, then grasp it firmly. As my hand reappears, I see that it is a young swift. Its body radiates warmth. I sense the staccato rhythm of its heartbeat.

I walk to the wide-open window, open my hand, and the bird vanishes like an arrow into the silvery light of the evening sky. Apparently he had flown down into the chimney and couldn't get back out.

Smiling, we look at one another. We can do nothing else—we have to smile. It is quiet, but then we hear the sound again. More rasping. I fumble about once more, reach deeper still into the cinders. I bring another swift into the light. I release it out the window to freedom where it arcs along the street between the houses and disappears around a bend. The swifts plunge into the air and are renewed to their element. In that moment, they are saved, even if they have already grown too weak to survive. I hear their exultant cries, their drawn-out shrills ringing in the evening. And the same exultation fills us, as we cannot help but fall into one another's arms, happy to be saved.

In that moment, I understand: The swifts are an element of the air, but they are also an element of happiness. The swifts are the children of the air's love of itself. And for the first time, I suspect that we do not properly understand this love when we limit it to an emotion only felt in moments when, for example, we try to keep a particularly desirable person in our life. On that summer evening, I have the impression that love is nothing more or less than pure aliveness in flesh and blood, with a beating heart and outspread wings. Indeed, that every moment in which we encounter life and its desires with affection, this love unfurls, just as those young birds unfurled their oversized wings into freedom and aliveness.

To love, I thought, means to be fully alive. But this has significant consequences. It became clear to me that this would cause us to completely rethink how we understand life and its significance. It would mean that, for some time, we might have understood very little—or have forgotten very much—about life and the feelings connected with it.

And it could be that the planet is not actually suffering from either an environmental crisis or an economic one. Instead, it could be that the Earth is currently suffering from a shortage of our love. As the planet has entered the sixth wave of extinction, more and more people complain about feelings of meaninglessness. Depression and personality disorders are on the rise, and billions of us continue to live in the gloom of abject poverty.

But this love—I thought, as the trail of the swift's arrowlike flight seemed burned like an afterimage into the evening's empty air—is indeed nothing other than the inexhaustible drive of both life-forms and the ecosystem to grow and to unfold. It is the desire for such unfolding and the joy experienced when that drive is fulfilled—and yet, it is also the happiness that my friend and I felt as we embraced one another after rescuing those little animals. It unfolds regardless of whether something good happens to me or to another being, because it is the joy experienced whenever life increases in the world, somewhere.

This love governs the certainty with which two cells find one another, the precision with which diligent molecules repair the fissures and breaks in our cells' DNA at the rate of a hundred thousand times per second. It accompanies the flowers' blooming and the beetles' instinctual feeding on the blossoms' pollen. Love directs all of these life processes—not as a kitschy feeling, but as an unbridled force with which creative energy molds the world into individuals and then destroys them.

Love, as I reflected on it, was something like the inside of aliveness. On that evening, I had slid into it—unintentionally, tentatively, a bit apprehensively—just as I had been shocked to encounter the pulsing warmth of that young bird in the rusty hollow of the stove.

And suddenly my heart beat just like that of the little swift. There were so many things to discover! There was so much that we could learn to understand about the extent to which we, too, participate in this force. And about how we could use it to become productively and poetically enlivened.

Eros: what makes the world more real

Whenever we talk about love, we usually think about people—couples above all; we think about togetherness and harmony, about disappointment and melodrama. We think about our love of ourselves, and not our love of the world. "Eros"—it sounds like a nice dessert, like a happy ending. But Eros, the Greek god of love, was considered a tragic figure in antiquity. He was not the god of pleasurable satisfaction, but of emotional intensity that burned just as hotly, if not more so, when unsatisfied.

But then, I reflected, weren't we—and with us, the great projects of our civilization—suffering from a great misapprehension? Had we collectively forgotten, perhaps, what was necessary to bring love into the world? That love needed to be called, kindled, nourished—and was not to be found and consumed like a commodity? Had we forgotten that it was not just pleasing ecstasy, but the benchmark for successful relationships in which more than one party—you and I, the individual and the world—could find common ground together? Striving for the most pleasant existence imaginable, for security, recognition, and regular nightly oblivion in this technologically comfortable and monstrous twenty-first century, hadn't we all become obsessed with an image of love that led us away from aliveness and sucked us down into a spiral of desires in whose middle was nothing but our own optimized selves—cut off, disconnected, and ultimately dead?

The search for love is a central movement that characterizes life, and it remains as inexhaustible as it is unfulfilled. Its consistent failure could mean that we understand love as little as we understand the living world, the natural world, the equally threatened creative power of the Earth. We often love incorrectly. This is the starting hypothesis of this book—in private and in politics, in big ways and in little ones, in our culture and in our beds. We love incorrectly because what we consider love does not enliven us, nor does it enliven the world. But is it at all possible to love "correctly"?

The answer that this book attempts to offer is this: To understand love, we must understand life. To be able to love, as subjects with feeling bodies, we must be able to be alive. To be allowed to be fully alive is to be loved. To allow oneself to be fully enlivened is to love oneself—and at the same time, to love the creative world, which is principally and profoundly alive. This is the fundamental thesis of erotic ecology.

Anyone who dismisses love cannot understand reality. This is true of all realities—physical reality and the reality of thought, but also especially the reality of the biosphere, the reality of bodies. No biological description is complete unless it is laid out as a biology of love. And conversely, we do not understand love if we fail to see that it is linked to the living world, to the experience of inhabiting a living body that trembles in joy and winces in pain. Love is a practice of enlivenment. The erotic is the genuine principle of life that permeates the world of bodies and life-forms.

I rediscovered this principle of life in the passion of the swifts. Those slender birds showed me what it is to be enraptured by existence. The swifts, who arced on sickle-shaped wings through the sky of Varese Ligure, through the cloudless air above that little town where I had my own apartment for a few short years. I experienced that love not only on the evening I rescued those two little birds; it surrounded me every day. It was an essence in the air. In the late afternoons, when the animals sliced through the atmosphere in flight and filled the sky with their shrill cries, it seemed to me as though I were actually immersed physically in their aliveness—as though, while running my errands to the baker and the tobacco store, I was walking through a substance made of pure joy.

In Italian, a swift is called *rondone*, with the emphasis on the second *o*. *Rondine*, with the emphasis on the first *o*, is the name of its little cousin, the swallow. Whistling and screeching, their bladelike wings carve circles into the balmy evening air above the market square. The birds already have the word *rondo*, which sounds like "round," there in their names. And even if etymologists suspect that these supple curves, the circles cut by their rapid flight, did not contribute anything to their naming, the ecstasy of that gyrating motion characterizes the phenomenon of the swifts.[2]

I remember that special evening in July when the birds—both the older and the newly fledged—filled the dull-blue and rose-pink sky above the town in swirling strata. It was a special evening because it struck me then how these birds suffused the air, dense as a cloud of solar dust and nimble as heated atoms. Perhaps because, for the first time that evening, the new flyers were feeling the sky as a lightness beneath their bodies' weight. The happiness of those young animals—"bird puppies," as my daughter always said—made the lifeless sky roar, enlivening the inorganic air.

There must have been hundreds of birds above the old, granite-gray castle in the village center. As though the evening had caused the air to thicken, as though an invisible reagent of transformation had turned the empty space into one trembling with inspiration and craving life. It was as though some chemical, a precipitating agent like phenolphthalein, had caused the air to thicken and crystallize to what it truly was—a solid shower of bodies eager to fully feel themselves alive. The twilit sky condensed into a thick cover of blended blues and reds, empirically attesting to the load-bearing capacity of the air. The birds, thus emerged, filled the sky with roaring curves, with buzzing arcs—their trajectories caused nothingness to thicken and to morph into flocks and fleets in passionate flight. I lay down on the edge of the village fountain and observed the sky's love for itself.

I cannot fly myself. What I am describing comes from my attempt to understand, with my body, the ecstasy of the swifts. I believe that this is a good method. Much escapes me, certainly—but I nevertheless understand the most important things. For I, too, am a creature with a feeling body who wants to experience joy and who must also someday die. I understand life because I am living. The South African Nobel Prize winner J. M. Coetzee has the heroine of his story "Elizabeth Costello" bravely claim, while at dinner with a number of philosophers skeptical about experience, that we know what another being—like a bat, for example—feels when it experiences itself at the height of its powers: "To be a living bat is to be full of being; being fully a bat is like being fully human, which is also to be full of being. Bat-being in the first case, human-being in the second, maybe; but those are secondary considerations. To be full of being is to live as a body-soul. One name for the experience of full being is *joy*."[3]

Joy is the sign of love; and love is the principle of a fulfilling equilibrium between the individual and the whole. The erotic manifests as that force that causes beings to inexhaustibly seek this equilibrium, to fail at that seeking, to neglect it, to temporarily achieve it. The power of the erotic suffuses the biosphere with life and imbues its members with the stamina to look—with new verve every day—for fruition, fulfillment, and joy. Conversely, the insistent striving solely for personal fulfillment in love amounts to an ecological tragedy. It follows the principle of taking from natural resources, rather than the principles of giving, sharing, and releasing.

Love as an ecological phenomenon

In the following chapters, I will trace the principle of life that creates a third through the contact of two poles—a relationship that transforms both sides. I will begin with the mineral world and then track, step by step, why every ecology—that is, every description of reality that understands it as an interconnected system of reciprocal inspiration, dependency, penetration, and the persistent search for freedom—centers on the principle of erotic attraction; and why every scientific description of the world that lacks this center ignores our central life experience. And since one can only write about love to the extent that one loves, these will be very subjective, deeply felt stories. I will speak more about this as well. Eros is the principle of creative plenitude, the principle of superfluity, of sharing, of communication, of the self-actualization that lies dormant even in rocks and minerals—the self-actualization that, as painful as it always is, cannot be avoided if we want to remain in contact with the reality of this world, whether as a thinker or as someone who simply *is*.

Nowadays, of course, this dimension is missing from most serious scientific descriptions of reality. The "king of all sciences," philosophy, has the concept *philia*, or love, in its name, but usually philosophical arguments sound more like complex jargon than an attempt to record encounters with the world marked by wonder, interest, and gratitude. And what about biology, our science of life, which many students continue to choose as their area of study because as children they felt an affectionate fascination with other beings? For the life sciences, the centrality of relationships and the significance of emotional experiences to ecological systems play only a minor role. Of course, biology is all about life-forms that can only exist because of relationships. But biologists typically describe these connections in the form of cause-and-effect chains. This approach has elevated biology to a leading science. But at its core, it is missing the dimension that could be understood to describe our experience of the world. In other words, the picture will only be completed when the biological description of reality is expanded to become an "ecology of love."

But this biology can no longer be the "mechanics of the heart," that the natural sciences still gladly offer up. It must be a practice of lived and fully experienced life, biology in the first person where every observation, every experience, every sentiment is put on the test bench of an image of reality. It must recognize that every being is in a state of ongoing transformation, constantly struggling to unfold itself.

Love is not a feeling, but the characteristic of a productive relationship. Failing to understand this is our great error in a time when all of us are chasing love as a life goal, but finding only an extraordinary lack of love, for which we then blame ourselves. Our misunderstanding of love continues to make this situation worse. A world in which love exists in fantasies but has no actual potency loses the ability to facilitate fair negotiations, bestow meaning, or produce anything other than purely monetary wealth.

Love is not a pleasant feeling, but the practical principle of creative enlivenment. This principle describes the way in which living communities on this planet—groups of cells, organisms, ecosystems, tribes, families—find their own identities while also fostering the relationships that they have with others and with the rest of the system surrounding them. Finding an equilibrium between one's own interests and those of others is the heart of love. In the experience of love, when my deepest desire is that my partner enjoy the greatest happiness, something more universal than a private feeling emerges. In that moment, love becomes a fundamental aspect of being alive. It is the sense of achievement in enlivened systems, where the freedom of the individual must always be harmonized with the whole.

Our stubborn insistence on the private enjoyment of a fulfilling relationship is, deep down, an ecological tragedy. For the idea of love as a resource that I need other people to give me reflects the view that the whole living world is a battleground of bounded estates and that evolution is the story of victors in a race toward optimization. This fits together with the idea that nothing is given—which is why you have to increase your market value (usually by increasing your physical attractiveness) in order to be lovable. On the other hand, an ecological view of love centers on other observations. It does not start with the idea that happiness can only be won in a fight and that to be clever, one should never give anything away. On the contrary, it believes that all essential things have already been given but can only be enjoyed if they are shared by all.

The affection of the body is mercy, not greed

A view of love as an ecological phenomenon is oriented toward relationships between life-forms in the biosphere. There, too, competition is only one side of reality. Putting the cascade of matter and existence into operation first requires a gift that asks for nothing in return: the sunlight given from the sky. The stability of a habitat is not guaranteed by the efforts of species and individuals to surpass one another. The logic of the living world relies much more on the fact that every species is dependent on another, that every act of taking is balanced by an act of giving. We have only just begun to understand how deeply this principle of the gift influences the world of organisms.

Predator and prey not only compete with one another in what biologists like to call the evolutionary "arms race"—they also become inextricably bound to one another. Tiny algae floating in the water have, over generations, evolved ever more complex body armor to fend off the crustaceans that feed on them. In response, the animals consuming them have developed ever more specialized mandibles. In the end, the predators come to depend completely on a particular form of prey, because they aren't able to feed on anything else.

The "predators" spare all other potential prey, for whom other forms of dominance and dependence are opened up, which further strengthens the interactions within the habitat and increases its diversity. The result is not something "better," but something "deeper": a greater degree of intimate interconnectivity. One could say that the entanglement of predators and prey in a common history leads to an intensive degree of interrelationship within the biosphere, which observers might then experience as that biosphere's beauty.

In erotic ecology, the feeling of joy is an integral component part of a flourishing ecosystem. Every relationship within the network of life produces meaning, because for the creatures concerned, it involves their whole lives—their existential desire to inhabit a body as a self and to continue unfolding that self. Through this experience—and this is precisely what makes it erotic—every creature can perceive its reflection in every other, because we all have a sensitive, vulnerable body that depends as much on bonds as on the air we breathe. According to this deep principle, we know how other beings feel, because they have bodies like we do.

The affection of this body is mercy, not greed.

In the evening, swarms of swifts arc through the air as it becomes transparent. The sky around the castle tower shrills with the birds' screeching. Hurtling, outpacing each other in squads that interpenetrate, intermingle, and then disperse again, they chase one other. The birds with their slender wings are a gift to the air. Farther above, at the edge of the sky, more flocks dance. The air has been filled with them as though with snowflakes, with dust, with sparks of sunlight. The swifts inhabit the air as sunlit froth, as though the old castle were a cliff, surrounded by the swell and surge of the sea. One animal after another plummets toward the walls and turns away in the last tenth of a second, leaning into the turn like a pilot flying a death-defying acrobatic maneuver. Or rather, the other way around: The pilot flying the maneuver turns like the playful swift. Our gaze moves skyward and does not turn away, seized by the birds' dynamism. Our necks bend to follow their curves, circles, and arcs, our eyes sucked upward into the chasing loop-the-loops and fleeing chicaneries of wind-taut bodies, nothing but wings, curving blades that cut tracks into the fabric of the sky. Speechless and humble as our limbs tingle with the joy of life, our gaze is an homage. The birds' infectious happiness is their trust in the air's capacity to carry them, the air's power to be void and thereby to support.

PART ONE

I

After the stone you fathom the rose.
After the rose you endure the stone.

CEES NOOTEBOOM[1]

— *chapter one* —

TOUCH

*The rivers flow not past but through us. Thrilling, tingling,
vibrating every fiber and cell of the substance of our
bodies, making them glide and sing.*

JOHN MUIR[1]

My second love story is about the rivers. It is a declaration of love to all that is earthly, to all that is not alive but from which we are made nonetheless: the limestone of the bones, the phosphorous of the genes, the iron of the blood, the salt of the tears, and the water that fills more than half of us. The story is about the smooth and effervescent streams, about the resting and the rolling, the virgin and the Creation-old stones over which the waters sweep. The story is about mountain brooks, wild downpours that flow in their beds of gravel and rock, about sparkling spouts that gush from tributaries, thoughtlessly wasting themselves as they follow the slanting terrain, whipping and stroking the stone with a patience that spans millennia in a single cascade.

The rivers carried us, swirled about us, and caressed us in those months of 2010, those long months from February till August, when I lived with my ten-year-old son in that tiny Ligurian town. The rivers guided us through the seasons, and they changed while we were changing. I let my hand glide over their icy pebbles in winter, and I immersed my whole body in the warm waters of calm stretches lying still in the summer sun. Often, the rivers felt like my nervous system. They slowly expanded through the landscape as though they were my own organs of perception. They carved their paths through the hills with a gesture that seemed to me like a nonchalant and aristocratic naturalness. The slow ways in which they decided their own destinies taught me the longing for my own aliveness.

Over and over, water showed me: The Eros of reality begins with touch. There is no life without contact. Without touch, there is no desire, no fulfillment—and no mind. When a light wave changes the structure of my retina, when I stroke the skin of my beloved, or when a nerve cell sends out an electrical impulse by spilling calcium ions, this is always an act of physical seizure. A physical seizure, no different from the coursing waters tossing and dragging the rolling pebbles against each other—the waters of the rivers, this purely inorganic world.

The Ligurian Apennines are a land of rivers. Not a region of quiet, wandering streams but one of swift waters that dart down, ringing, into the valleys, causing the mountains to crumble in shards and reassembling them into new configurations. It is a land of wild mountain brooks that crack crags into pebbles and thereby think up the topography a second time, composing it anew out of boulders, blocks, slivers, marbles, sand. Liguria is the land of rivers that are creation—creation in the form of an ongoing event, Creation-old creation—rivers that turn all the ground they touch into a geological nursling and call forth eternal youth. In the rivers of Liguria that bedew our hands with the cold moisture of their waters and refresh them with the warm weight of their never-settling pebbles, we are in the childhood of the world.

Indiscernibly, the rugged hills with their velvety cover of oaks, ash, cherries, chestnuts, and hazelnuts, are ruled by the rivers. The rivers' stony pathways flow back to the primal source. Those who live close to those rocky rivers, running through their valleys, will always be reminded that there was a beginning that no human planned or designed, a beginning in which humans and rolling stones were still one and the same. Pebbles, tumbled together, outspread and arranged by nothingness, devoted gravel paths, white corridors in the ravines that widen and narrow again, divide and embrace one another once more, like snowfields gentle, a stony flurry of flecks at the base of the valleys.

In the rivers, the striped stones roll, just as they do on the beach. The rivers have filled the ocean to the brim with their rolling cascades. Wherever their waters run, I see myself—in the swift sparkle, the surf's breakers. The same message shines to me from all waters. Water is a mirror made of a thousand mirrors. I cascade down with the rivers' scree, I breathe myself out in the beaches' pebbles, I am the undulating world that rhythmically rises and smoothly dies away. Everything has already happened. Yesterday, today, and tomorrow collapse together in the quivering line of light across the stone, on the epidermis of a bud of granite.

Gravity, erotic power

Rivers form because moist air rises above the mountains and the water vapor within it cools, condenses, and falls as rain that runs downhill. The valley attracts the water's weightiness. Rivers are thus an indirect and far-reaching phenomenon of gravity. To put it more poetically: Rivers, their ravines, their full-fledged valleys, the gravel beds—these are all the ways in which the mountains perceive gravity and reveal this perception at the same time. Of course, mountains perceive nothing in the same way we do, nor even in the same way bacteria do. But they change because of the waters' touch, and we can see and sense this alteration. Aldo Leopold called the perception of this transformation, "thinking like a mountain" (and suggested that we might be able to solve a significant number of environmental problems by means of such a different way of thinking): not fabricating rational correlations, but rather following the traces that touch leaves behind and recognizing therein the fact of constant transformation.

Thinking like a mountain also means coming to understand that on this planet there is a foundational erotic attraction between all bodies, a pull that calls me, my body, toward others just as the valley attracts the waters. Larger bodies attract smaller ones—the sun attracts the Earth, the Earth the waters. Gravitation, according to the perspective of Aldo Leopold's thinking mountain, is the Earth's tender longing for us. Most of the time, we no longer perceive the Earth's maternal pull. But is it not partly because of this tenderness that we feel comforted by lying down on a meadow during hard moments, bedding down our bodies on the Earth?

For the American ecopsychologist David Abram, the foundational moment of every relationship to the world, based in an erotic bond, is established by the circumstance that my body is always pulled toward the larger body of the planet. And gravitation is only one of the fundamental forces in the universe. There are three other principles of material interaction. Already in the name of this concept—interaction—physics signals that we are dealing here with phenomena of exchange, of relationship, and of contact. In addition to gravity, the "standard model" of physics cites electromagnetic force (which includes magnetism, charges, and fields) as well as the weak and the strong nuclear forces that hold the interior of the atom together.

All of these fundamental physical laws formulate the principles of relationship between material particles. They predict what will result whenever diverse agents enter into a scuffle with one another. Nonetheless, natural science does not attribute much significance to this aspect of the incessant jumble at the heart of matter. It sees the principles, but it does not take the phenomena very seriously. For natural science, the four fundamental forces amount to a set of laws that one must take into account in order to analytically understand and rationally reshape the cosmos. Few physicists also see the fundamental principles of contact and penetration manifesting through these forces.

Of course, physics is in the right—though not completely and entirely. The mathematical structure of the object that high schoolers try so arduously to fix in their brains is always also a cabinet of possibilities for mutual transformation. When the universe came into being out of the Big Bang, there was basically nothing except this possibility: the energy of beginning and the force that variously gathered together or tore apart an unformed Something still concealed within it as potential. Everything after that resulted from matter's eternal readiness to form new attachments.

Natalie Knapp, a Berlin-based philosopher, views the world as made up of multilayered bonds, rather than solid, congealed things or victorious, solitary survivors. She thinks that atoms feel something like desire for one another, a desire to be more, to transform themselves into collectively constructed, complex molecules. For Knapp, this is an elementary act of love.[2]

No matter how we describe it—whether we are sympathetic to such an understanding of cosmic tenderness or would rather document evolution and the slow increase in complexity with scientific neutrality—in either case, we can maintain that in this world, there is something like an ever-active tendency in things to congregate and to bond together into new, more complex, more sophisticated forms. For this reason, we could also describe physics as a science of relationships.[3]

Since all contact between bodies in this world calls forth new possibilities of meaning, the complexity of the world can increase even though energy in the universe constantly seeks equilibrium. The Eros of matter counterbalances the physicists' basic assumption that "entropy" in the universe is constantly increasing, meaning that everything in the cosmos is trending toward a uniform condition of the lowest thinkable level of energy. Fires burn out. Life-forms die. Our bodies break down. Even the sun will collapse someday.

Although everything is, on balance, moving toward becoming a uniform expanse of sameness like the countless boring grains of sand spread across the Sahara, the whole process is enabled solely by the fact that knots and chains are forming everywhere as it unfolds, and in those spots, complications arise. One can imagine this phenomenon as something like the fuzzy eddies that form in a glass of clear water when its contents are mixed with fruit syrup: Everywhere in the world there are places of encounter where substances react, bonds form, electrons jump from their shells into those of other molecules, where massy bodies begin to attract one another, so from the stories of individual participants, new situations emerge that are always brought forth mutually by all the participants.

Life is the clearest example of the cosmos's poignant and dramatic striving toward a future point of rest predicted by physical laws, while also taking countless detours along the way. As the energy level of material particles sinks overall, liberated forces of propulsion are used to spin out new relationships, to arrange unprecedented constellations, to create never-before-seen forms, smells, and experiences. As the universe heads unavoidably toward what the physicists call "heat death"—that fade-out into solitude and uniformity—life freeloads aboard this gradual cool-down process to spawn its own oddities: coral reefs, little rotifers, naked mole rats, you and me.

In the face of failure, the material world has a constant and innate tendency to create anew, to organize new forms. It could simply collapse into heat death, but instead it invents more elaborations and playful arabesques: Energy clumps into atoms, atoms pull together as molecules, molecules assemble to form chain reactions, and these eventually enclose themselves as living cells against the environment that produced them. While the stillness of death looms and tends to consume all available energy, unprecedented new relationships between elements form and characteristics emerge that atoms had not even dreamt of previously.

The biologist Stuart Kauffman delivered several pioneering insights into this very subject. It is evident to him that complex forms will emerge necessarily from any heap of unorganized matter if we only wait long enough—and that, at some point, these forms would develop lives of their own: that tenacious will to live that we all know in the deepest parts of ourselves. The key word with which Kauffman describes this process is "autocatalysis": mutual aid. Kauffman observed that once a sufficient number of reactive substance classes were present in a primeval soup, it was only a matter of time before they began to help one another to produce new, unforeseen bonds.

The more varied the mixture, the greater the probability that the spark will ignite, causing each of the various molecules to help produce another, thereby completing the circle: The whole begins to stabilize itself as a whole. The individual components help one another to mutually overcome the downward trend. They defy their fates together by constantly forming new relationships, which in turn produce new substances that help to further postpone their collective demise.

We, too, are a result of this dance: The cells that make us up are direct descendants of the first autocatalytic chains. Every life-form is the successor of an unbroken catenation of life reaching back to the earliest forms of self-organization. We bear, within each and every one of our molecules, the triumph over heat death—a piece of an exploding star.

Contact forms the scaffolding of reality

The deeper physics penetrates into the innermost areas of matter, the clearer it becomes that no material certainties remain within it; yet everything there can instead be understood in the form of relationships. The nucleus of the atom with electrons moving around it is not, in fact, a tiny planet with orbiting satellites. This is simply a useful model, a translation into a worldview of individual, clearly distinct things that is easier for us to comprehend. In truth, the atom is a relationship between different likelihoods of energy concentration. Its shape at any given moment is essentially the snapshot of a particular configuration of these relations.

The distribution of electrons around the nucleus is described by the so-called Schrödinger equation. But this equation does not map a topography like the Diercke Atlas or Google Maps; instead, it specifies probabilities regarding the most likely locations of the particular atomic elements. It would perhaps be most apt to say that the equation describes how an atom "feels in space." The Schrödinger equation models the possibility of a certain relationship and then formulates the odds for the various ways in which that relationship could be lived out. Contemporary physicists still debate how to make sense of that mysterious equation. At its center stands the fact that all matter, including myself, can only be understood as the experience of being in relationship.

Through its ongoing push into the world of the smallest particles, physics had eventually reached a level that showed that the cleanly divided units that we had habitually considered to be the constituent parts of the world were simply different facets of a common relational nexus at their cores. But we do not need to be quantum physicists to understand this. That is the exciting thing. All we need is a path to the outdoors. All we need is a walk by a swiftly flowing river. All we need is to inhale deeply, letting the surroundings fill our lungs.

We are still at the beginning of the world.

In that sense, then as now, every alpine brook in Liguria flows down into the valley with the unbridled energy of the Big Bang. In this world of stones, of water molecules, of minerals that polish one another as they slowly wander down into the valleys, tugged along by gravity—in this world we discern the fundamental principles of erotic touch: Two sides always enter into relationship such that both come away changed. The river gravel is stone that water has transformed into a flowing form, and the swiftly cascading water is liquid that the stone has shattered and cracked. Only together do they reveal a meaning, only by altering one another do they become what they are—something much more than what they were before. The hard, crystallized granite, polished by centuries, discovers within itself the potential for curving and flowing, and the supple water experiences its muscular, voluminous cataracts as its share in the quality of massiveness.

This world is not populated by singular, autonomous, sovereign beings. It comprises a constantly oscillating network of dynamic interactions in which one thing changes through the change of another. The relationship counts, not the substance. And to make this relationship possible, it is necessary that the two sides touch each other, that they nestle into one another, penetrate one another, grind themselves against each other. This is the fundamental erotic that constantly makes new things out of other things. By coming into contact with the water, the stone's form softens, and through that softening, it begins to show characteristics that are in fundamental opposition to itself. The water crashing into the valley strikes the blocks with almost rigid force, and therein gains a quality diametrically opposed to its fluidity.

Both of the parties who engage in erotic contact are not only transformed into something other than what they were before; erotic contact has the potential to transform them into the very opposite of what they had been before. The stone in the riverbed truly comes into its own by getting mixed up with its antithesis, the flowing water. It gains its distinctiveness as river pebble by losing its self-identity and becoming something else. A piece of productive transformation. Not constancy but the outcome of a creative exchange.

This exchange, which draws upon the contradictory forces of various bodies and leaves nothing as it was before, is the basis for the principles of an erotic ecology. It professes the foundational rules of the physical cosmos, which we share with everything else—with bats and ferns and rotifers and viruses and crystals. And this exchange between bodies, which causes constant transformation and leaves nothing as it is, also determines the terms of life.

So the first approximation of the axioms of an erotic worldview might look something like this:

1. The world comprises matter, bodies that are in continuous contact with one another.

2. The contact between bodies forms the scaffolding of reality. Reality is physical and not abstract or "ultimately mind." Reality is not neutral, because every instance of contact leaves behind irreversible traces that change all of the parties concerned.

3. Because of these moments of contact, a layer of meaning is called forth: Touch is an alteration with meaningful results. Moments of contact cause relationships and, for living beings, interests. Thanks to a purely external circumstance—the fact that things are pulled toward one another—a "plane of inner experience" is formed.

4. Being in contact has a fundamentally positive, almost desirous aspect: The individual parts of the world are drawn to one another, attracted to one another, one might say interested in one another. Even if the moment of contact is incidental, the results of it are not. Contact is unavoidable, as is the resultant inner interweaving of the network of relationships that makes up reality.

5. Moments of contact, interweavings, new relationships, and complications are ever open to the possibility of being carried further. Reality itself constantly facilitates the development and proper expression of new erotic relationships. The cosmos manifests a gigantic process of "autocatalysis" in which relationships—structures and meaning—come into being and facilitate further relationships and deeper experiences of meaning.

6. Everything that is of this world longs for more contact in order to be more strongly and intimately in relationship, and thereby to come more fully into its own.

"Water transforms our mirror image into nature"[4]

I discover my very own Ligurian brook on the third day after my arrival. I find the brook, and immediately it belongs to me. Not in the exclusive sense of our economics; I do not want to take anything for myself—on the contrary. Every pebble shall be allowed to lie wherever the elements have washed it, every gleaming micaceous fleck shall rest wherever it fell in bygone days under an inaudible crash of thunder. The brook belongs to me just as certain emotions, preserved from childhood, are entirely my own, just as certain gestures, certain types of facial expressions make up a portion of my personality. It seems to me as though this is the water on whose banks I have always perched as I aged. "Water is the gaze of the Earth, its instrument for looking at time," writes the French writer Paul Claudel.

It is still winter. On this dark day I drive, aimless and confused, up the tarred roads that wind into the rugged, slush-soaked and sodden hills. Finally, I follow a weather-beaten sign that points the way toward Porciorasco. The drives dive into the underbrush, curve along the cliffs, climb the hillcrests. The icy patches that I have to pass gleam dully. I drive over a bridge of slender stone archways, water foaming beneath it and making its blocks rumble hollowly. The river forms the base of a valley that widens, gray and blue, into the Mediterranean. No habitations are to be found, no trace of humanity.

Porciorasco is a ghost town. There are yawning gaps in the outer wall of its baroque church. A village square of mud in which centuries-old holly oak trees grow: dark green, unapproachable. Almost all the buildings, ruins. Many of them have been converted to barns and the smell of cow dung wafts from them. Behind the village, the street becomes a very narrow lane. I park the car and follow it on foot down into the valley. Coldness has dug its claws into the slope, hoarfrost glistens on sharp-edged stones. The lane ends at a ford—evidently the farmer drives onto this road through the clear waters. The brook comes shooting out in a curve from a small valley. Alders and ashes point with thin, bare, outstretched branches toward the sky, currently colored pale blue. Goldcrests chirp—quiet, airy, a delicate shower of crystal. The brook bounds almost soundlessly from block to block, round stones, big as ovens, as small cars, meteorites fallen out of time. I balance on them facing upstream.

Unwilling to be checked by the cold, pussy willows hang over the banks of the brook, which is the gray-green of glacial melt. The water bubbles over a branch lying across the stones. It has formed thick icicles in an organ-pipe array. The water plays with the rocks, the frost plays with the water. At a clear patch, I toss in a small pebble. The plop resounds like a cannon shot in the stony valley.

On a misty winter morning a few days later, as clouds settle over the hillsides, I see the water ouzel on the stones of the brook by the old bridge in the middle of the village. From a distance, it is nothing more than a little black speck on the rocks that vanishes into the foam of the alpine brook in the blink of an eye. The icicles that hang on the siliceous blocks and the fallen wood over the sparkling water, the brown and gray and white of the little bird's feathers—weave together into a figure of calm and cold and stoniness that comprises this animal's world. Aha, I think. This is how it is. The ouzel is this rocky chill; it is the chill's inner side. It is its aliveness.

Bobbing delicately as it lands on the granite after every brief flight, its white bib as pristine as the foam of the virgin brook, this little bird seemed—caught up as it was in the laws of flow and compelled to tend to its own life (and to those of its offspring)—to be something like a result of the water. It carried on in a largely easygoing manner, whereas the cold into which it repeatedly dove, bedecked in its ball gown of feathers, had become unbearable to me. The bird acted like water following the pull of gravity, and like the frothing waters that seemed so eager to cross to the valley through the rapids of a particularly narrow passage, perhaps the bird with its fragile body was just as smitten with the prospect of wandering along the ground and over the stones, impaling larvae with its beak.

The water ouzel simply did what it did, and yet it appeared, in so doing, to shine, and to be something special. It formed one of the necessary, but also arbitrary, knots of creation, which often appear to us as gifts haphazardly disseminated, despite their inevitability. But these gifts only prove that this exchange of contact, reflection, and transformation has no other goal than tenderness: being here in a state of distinctive particularity and dutiful rapture.

John Muir wrote in his notebook: "How interesting it would be to keep close beside an ouzel all his life, and be present at his death-bed! Surely there would be no gloom, no pain. I fancy he would vanish like a flower, or a foam-bell at the foot of a waterfall."[5] Muir hit upon something crucial when he described the water ouzel as the "hummingbird of the blooming water" at some later point in these reflections. He understood that the same aliveness courses through all beings, that their vivacity is also mine.

The idea of being cut off from the other, as laboratory researchers are from their objects of inquiry, is perhaps the fundamental error of our civilization. The delusion inherent in this idea is what makes possible our unbelievable indifference to the widespread death of the natural world. Muir saw that the water ouzel *is* the rocky chill of the wild brooks, that it is its inner side. He saw that the water ouzel is its aliveness. And that the small bird's abandoned work in the cold foam of the alpine brooks makes me alive, because it can fill me with love.

I walk farther, as the suddenly failing light extinguishes the scree of the wild brook. The rivers of this landscape truly flow through me. They form an extension of my nervous system. They teach me to see.

The center of reality: all things answer one another

During the months in Northern Italy, my soul spent its most important moments by the rivers, by little rills that hopped fleet-footed over round clods, by cool hollows under the shadow of branches that screened the shimmering midday glare, by the far-reaching combes of wild, broad, turquoise rivers. I am glad that my son learned the beauty of rivers there.

Once, in the summer, we stopped behind a ledge that the water had curved around; the bygone flood had flung a tree onto its rocky tip. We climbed down over the porous rocks into the riverbed of shiny pebbles. Max broke a few pieces out of a vein of crystals and began to rinse them in the brook, coloring the shallow waters at the edge with the milky white of the stone. Aimlessly, I sought out the most beautiful pieces—round, lightly porous pebbles; flat, smooth pieces with brown grain and thin white lines, upon which I resolved to write, in a playful hand, a letter to my daughter, whom I dearly missed.

There were gemstones, precious stones everywhere—one needed only to bend over and pick them up. The world was overflowing with treasure. I quickly gathered far more than I could carry. I would have taken home the entire stone-filled riverbed if I could have. I had enough stones for an entire library of letters to my daughter. "Why don't I just set myself up here?" I asked myself. Why cram my pockets full of this beauty and nearness—why not do it the other way around: Come here, stay? We humans are such twisted beings, void of memory. The pebbles in the river, the oldest pieces of the Earth's crust, the minerals that make up my insides: Their cold, velvety-cold touch was tenderer than a kiss.

The effects of water bring to mind the conditions of the omnipresent physical Eros that generates incalculable new interpretations of existence through processes of transformation and physical penetration. The stony beds of the mountain streams present a reflection of the act of flowing, a commentary on the abstract phenomenon of running water that trans-forms its litheness and alacrity into the language of rock. Water leaves behind traces on fixed objects that reflect its fluid nature. It provokes a question-answer game of the elements. Water that runs over sand—at low tide, for example—carves characteristic branching patterns that fan out across the ground. The traces of water look like water, though they are something different. They are a reflection of water in another medium, and this makes them also into the opposite of water: into something fixed. Cliffs, stones, sand.

From up close, such a pattern of water channels on a few square meters of tidal beach, or on a sandy forest path after the rain, looks exactly like a river delta seen from the air. Seen from above, the Mississippi is ulti-mately nothing more than water running through an oversized sandpit. This phenomenon, which the mathematicians call "fractal growth," tells us that big things are just enlarged reflections of small things. Nothing is left without traces. And even when no one perceives it and when it seems to make no difference to any living being, every single event—a material condition, a change in phase, a gust of wind, a withering heat wave, an abrasive surge of sand—all of these things are experienced as changes by other bodies. Every touch leaves behind traces that contain something of both participants, because it reflects in part the force of the impingement and in part the compliance of the touched.

In the 1950s, the Russian psychologist Sergei L. Rubinstein described this foundational, erotic state of affairs as a "universal phenomenon of reflection." He even attempted to found his own theory of perception based on it.[6] Reflection means that nothing else is left to the objects of this world than to establish connections between one another and to present the impressions that result. Nothing happens with ulterior motives—and yet, nothing remains without traces on it. "The wild geese do not intend to cast their reflection. The water has no mind to receive the image."—this is how a Zen-Koan articulates the unintentional poetry of reality.[7] But once touched by the image of the wild goose, the quality of the water—an inanimate fluid—is changed; it becomes an imaginative substance.

Such forms of intentionless exchange between larger or smaller bodies can establish new developmental branches of the biosphere. The gravity of the moon, for instance, pulls at the Earth without meaning to, causing our planet to arch toward it like the body of a lover such that oceans rise up and simultaneously retract their waters from the coasts of the Earth at low tide. One could say that tides are the way the Earth perceives—or "thinks"—the moon. And this "thinking," which is of course not an active experience, nor is it in any way guided by conscious interest—though it is still a kind of (unintentional) answer—results in an entire universe of unique consequences: It lays the foundation for the magnificent universe of animals and plants, the muscles, crabs, worms, coral, algae, and aquatic plants, that have gained a foothold in the ocean's tidal areas. So one could say that the Earth's perception of the moon enriches it by establishing new niches, new opportunities for aliveness. Because the moon excites its imagination, the Earth evolves new possibilities for self-transformation.

This constant babble of call and response pervades our geo- and biosphere. We are everywhere confronted by the erotics of the encounter. The blooming hedgerow in May—that effervescent tidal foam of hawthorn and clematis between the shady trees and the bright meadow—is the product of such erotic contact. The hedgerow at the edge of the woods results from the encounter of the dense forest and the clear open countryside. The blooming hedgerow shows how the woods answer the meadows. Or how the countryside dreams the woods: hence the flowers. And at the same time, the hedgerow is neither forest nor meadow; it is something wholly unto itself in which both forest and meadow are extinguished and reemerge, altered and combined.

To describe this poetry of traces that leave a legacy of further traces, the Chilean cognitive scientist, biophilosopher, and Buddhist Francisco J. Varela used the complicated-sounding concept of "reciprocal specification"—an act of mutual engendering. Only through a moment of encounter does one's own character come fully to fruition. The world is not an aggregation of things, but rather a symphony of relationships between many participants that are altered by the interaction: a necessarily erotic occurrence.

Such a perspective can resolve the scholars' old debate: What comes first, our notions of things or their "objective" qualities? Philosophically speaking: Is the world empirically real? Or have we simply constructed all of it? Do the "facts" determine our thinking? Or does society? Or does language, the so-called "discourse"? In a cosmos of erotic reciprocity, such oppositions have no meaning. For relationships always occur between two extremes that they also contain, and in such a relationship of reciprocal interpenetration, neither extreme remains as it was prior to the encounter. Varela's colleague, who died some years before him, the cognitive scientist and anthropologist Gregory Bateson, often said that everything we perceive is a difference that has a meaningful effect—"a difference that makes a difference." And such a difference is always the result of a relationship, which is to say, of an erotic exchange.[8]

Turning our attention to the sense functions that enable perception in the first place, we find that the erotic phenomenon of contact, or physical embrace and penetration, is in effect here as well. We are only able to see because the energy of a photon bends around the protein stacks of our photoreceptors. In order to pass the stimulus along, channels for charged particles must be established in the cell membrane. The energy for this is provided by the transformation of substances that we put into our mouths and chew. Perceiving warmth means that the physical forms of the antennae of certain sensory cells have been deformed by infrared waves. In order to read the data found in DNA, their strings must be channeled through complex formations of protein. When an animal smells a fellow member of its species, the scent causes a physical deformation of certain areas of the sensory cells in the nose. Birdsong makes the air vibrate, which in turn sets perceptive membranes like our eardrums in motion, which subsequently transfers the vibrations to incredibly fine, sensitive hairs rooted in fluid, which finally pass on the stimulus through the excitation of electrically charged atoms. All this is touch.

All of these are examples of physical encounters, without which no abstract information could be processed in the biosphere. All of these are examples of how deeply we are enclosed within a cosmos of contact and caresses. The Veronese Renaissance doctor and scholar Girolamo Fracastoro, who was the first to formulate a modern theory of infection through pathogenic germs, used the designation "universal attraction" to describe this Eros—"*sympathia universalis.*"[9] Even in the smallest atomic particles, every act of existing is a meeting and a desire for encounter, and thus all acts of existence are genuinely open to the outside, deeply creative, and intensely imaginative.

Afterward, it is never as it was before

On the banks of the little Ligurian brooks in the summer, one feels another impulse of this foundational erotics, this sympathia universalis, a unique and particular mirror that casts its light upon us. The things cry out: "Do as we do!" "Immerse yourself in this sympathy!" "Touch and let yourself be touched!" "Be naked like us, the elements!"

Waiting cold and untouched under the sunbaked alder branches of insatiate summer days, the still, reflective surfaces of the wordlessly calm water are an invitation to the skin, an invitation to bare oneself and descend, an enticing temptation to assent to the drive for touch.

I know a brook that runs along a narrow green valley, aspen-shaded and blackberry-veiled, hurrying translucent and cold over granite hearts and gravel slack—a brook that has always seemed to me like the model for all brooks: that model of a river that God keeps in a secret workshop, tucked away in the drawer labeled "brook," the one used to form all the other rills and streams. Its water comes flowing straight out of the childhood of the world.

And this is where I did not ask you to undress, to hastily discard your clothes; under the summer's late shimmer, under the mute midday glare, this is where there was no more place for questions because the water itself was longing for touch and the sweet pain of the cold made the skin into an organ of shock, a shiver as we realized how large a moment can be. We doffed our clothes, plunged into that first, archetypal river, and after the cold exhilaration we embraced on the warm stones until, for a few moments, things were no longer distinct from one another, until the boundaries between the trembling cold of the water and the warming solidity of the stones dissolved into an acquiescence beyond pleasure and pain, into a form of sensory observation that inwardly noted every detail despite being in a state of utter rapture, a unification in the mineral heart. Ecstasy. Equanimity.

Thinking like a river above which the aspens in the silent heat stand and will go on standing. Thinking like a newt, like a little bird. A lichen. It is only a matter of imitating the river and transforming one another through physical contact. Your skin, white despite the summer's suntan, born anew in the water; it was the place that I had wanted to take you all along, the place that I knew you would understand because you sensed as well as I the extent to which our encounter has changed us. It was the place that I wanted to gift to you, and you accepted it and understood it and gifted your skin to the water, and the water laid you in my arms—completely new and just as you were originally imagined to be.

— *chapter two* —

DESIRE

For what else have stones been shaped
but to prolong the human presence and to say
soundlessly in lost tongues:
We loved the earth
but could not stay?

LOREN EISELEY[1]

Smoke hangs in the cold air, the aroma of burning oak and cherry logs. The last of winter is trapped in the cliffs and the trees and the weathered pastel house stones, just as a long fever lingers in a convalescent's bones. The first rosy hues of returning life color the hillsides from the west and leave a pinkish sheen on the thick river pebbles arranged along the banks and hollows. Voices sound from the trees by the Vara river, from the bare lindens with their long fingerlike branches, from the one tall cedar in the park. A blackbird calls, a song thrush answers with a few choppy, repetitious cries.

But at the top of the cedar, undeterred, the mistle thrush, that most melancholy of all song birds, delivers its monotonous spring tidings. The valley fills with sound, just as the blue and crystal-gray of the granite has begun to blush today with a hint of summer. It is that evening—one such evening comes every year—when all the promises of newly blooming life have assembled. Nothing is fulfilled. But everything seems possible. The mistle thrush's song is part of this promise, and perhaps its only possible fulfillment. The mistle thrush is a creature very distant from me, and yet my pre-spring world comes from it. We are both necessary: Spring is created through both of us together.

Touch and material permeation are the foundational forces of reality. In the last chapter, I described how they govern the realm of physics, the behavior of atoms, of molecules, of water, of crystals, and of stones. But to an even greater degree this erotic exchange characterizes what is happening in the biosphere. Mutual transformations determine the living world and every experience of the flesh and of the senses, and thereby all instances of sensory experience and all understanding of meaning. Biology is the science of that contact from which all experiences of touch emerge, the theory of interchange in an endless variety of forms, of metabolism—the exchange and transformation of matter—of amalgamation and of parasitization. Biology is the erotic science *par excellence*, because a living being is an erotic process: It transforms itself through contact with others, imagining new relationships out of each existing one, desiring more life, unrelentingly seeking a connection to the whole of which it is both concentration and unrepeatable instantiation.

The erotic within the living world can be most clearly seen in three domains: the spheres of symbiosis, communication, and metabolism. The permeation of different bodies is extensive enough in each case that when you look at it deeply, it is not easy to make out clear boundaries around distinct organisms. Varela spoke about living beings as "selves without a self."[2] A being comprises organs, which are made up of tissues, which are made up of cells, all of which have a certain independence. Add to that the symbiotic organisms like our intestinal bacteria, which we could not imagine living without. "Species purity" is thus an illusion. Organisms are already ecosystems operating within ecosystems. The fundamental objective of this arrangement is to transform sunlight into flesh and flesh into other flesh: a completely earthly, material sharing of the same substance.

Our sympathy for the natural world is already, at the bodily level, an erotic attraction. Privately, we know (even if scientific culture has been trying for some time to teach us otherwise) that we are an ineluctable part of a material exchange, at the crossroads of an ongoing turnover of productive energy that generates our intensive experience of being worldly subjects, existentially affected by all of these things. Because our whole being takes part in this exchange, and because its success or failure determines our fate, we *feel*. We unfold our deeply sensitive and expressively poetic existence as a feeling part of an organic whole.

This understanding of life as an interconnected network of relationships, in which each being is simultaneously center and periphery and which is so close to our simple lived experience, has been relegated to the status of a romantic dream for many hundreds of years. In our scientific picture of the world, we have gotten very far away from this. In our daily life experience, on the other hand, this form of perception determines the extent to which we can be ourselves. It is still the center.

A lesson from the fireflies

Anyone who believes that life is a battlefield full of individual warriors should go out into the meadows on a spring night. There, you can learn that the biosphere does not spawn cutoff, clearly differentiated individuals who compete against one another—assuming you find such a meadow; that is, now that some farmers have started to sow a single, standardized species of grass.

In my little Italian village, the narrow streets climb into the hills where the meadows are still allowed to grow wild in the springtime. Within two or three weeks, the stalks swell into a multitude of meadow-grasses and blossoms as tall as my waist, fragrant and enveloping. I think then: It might have been this way once, when the plenitude of existence could spread everywhere and it seemed unavoidable that every corner of this biosphere would fill to the brim with life. When it was only natural to think of this cosmos as living, as enlivened at its deepest core, and not as an optimized assemblage of dead matter. If you want to understand the extent to which your own existence results from the collective work of diverse organisms, you must go outside on nights such as this, when the moonlight on the diaphanous hillsides makes them seem almost translucent and fireflies tumble through the gloaming like tiny stars gone astray. Yes, it is still out there, even in Europe, if you go looking for it.

Such an experience of the harmony between a landscape and its life-forms is probably not the result of objective analysis. But this is precisely the point: If you let the calyxes and grasses slide through your hands amid the firefly flurries, celebrating the coming summer, you don't just perceive a multitude of other beings—the hundred or so species of plants and countless insects that make up the meadow's ecosystem. You also experience yourself as a part of this scene. And *this* is probably the most powerful effect of experiences in the natural world. When you immerse yourself in the natural world, you wander a little through the landscape of your soul.

For a long time now, such experiences have been considered not very reliable, certainly unscientific, and, if valid at all, deeply steeped in that pleasant state of mind known to us from fairy tales, novels, and poems. The moonlit night, for sure. Eichendorff! Is this supposedly where "the sky had silently kissed the earth"?[3] And yet, in the interplay of the meadow's plants, insects, and microorganisms, and in the night wanderer's experience of this interplay and of his partaking in it, those familiar with recent biological research cannot fail to see a clearly tangible example of the principles upon which the world of life-forms is based. Seen in this light, the night wanderer's sense of belonging, of deep investment, is not a fallacy, but stands at the center of a realistic experience of what is actually meant by aliveness. Not theoretically, but practically, experienced from inside of a living being, which is what we are.

The principles made apparent by biological research show us that life is, at nearly every level, a collective concern, a shared enterprise undertaken by a wide variety of beings that arrive at a stable, functional, and thereby beautiful ecosystem by somehow putting up with one another and reaching agreements. Rivalry, competition, and selection in the Darwinian sense definitely play a role, but this is not the merciless final word; it is simply one force among many that living systems use to create and form themselves out of a multiplicity of participants. "Symbiosis" is the term often used for this cooperative process. But "symbiosis" has an overly pleasant ring to it that suppresses the fact that an ecosystem's success produces not only the happiness of brotherhood but also the horrors of annihilation. Eating others and being eaten (which lies ahead for all of us) figure into the same living fabric, as processes necessary to maintaining the stability of the whole and allowing it to experience itself.

For that reason, it would be better to say that biologists understand that life is a phenomenon of absolute communality. Flourishing in a relationship of mutual benefit is as much a part of this as lustily consuming another in order to guarantee one's own flourishing. The most astonishing thing about a meadow is not only the fact that the plants growing there create niches and a mutually beneficial micro-climate, but also that the stalks of those same plants have to be grazed in order for the meadow to remain a meadow. Their leaves and buds must be shredded by the mandibles of countless insects, to be crushed by rabbits, deer, and cows, so that they might perennially reemerge, variegated and placid.

The biosphere is full of such transformations. It is the continual product of them. There is no being, no life circumstance that does not result from contact, penetration, and conversion. The cells of our bodies result from "endosymbiosis," from the contact between two different types of bacterial cells in which one of the cell types encloses the other. Only by this transformation into the body part of an other could the enclosed bacteria further evolve into the organs necessary for the life of the enclosing cell. By infecting us throughout the course of our phylogeny, a multitude of viruses have infiltrated our genetic material with their DNA. The function of that DNA has transformed within our genetic matter such that it has become an indispensable part of our bodily processes. The living world is a constant conversion of one thing into another, leading to inexorable new growth.

In its incessantly renewing plenitude of life, the biosphere is no more "truthfully 'symbiotic'" than it is "fundamentally 'competitive.'" There is only one immutable truth: No being is purely individual; nothing comprises only itself. Everything is composed of foreign cells, foreign symbionts, foreign thoughts. This makes each life-form less like an individual warrior and more like a tiny universe, tumbling extravagantly through life like the fireflies orbiting one in the night. Being alive means participating in permanent community and continually reinventing oneself as part of an immeasurable network of relationships. This life network is knotted to all individuals. But just a single pull, a single slipup, is enough to loosen the ties.

If you walk through the evening meadow you experience all of this in a mysterious manner. Relief washes over you, because all at once your own struggle for life, the demand to somehow get yourself through the days, is poignantly and reassuringly echoed back to you. The burden is carried by an aliveness that vibrates everywhere, elevated above such troubles for the span of a springtime evening. If you hear the quiet rustle of the wind in the grass, you recall, deep within your body, that you are not a solid, constant individual. The grass seeds, scattered carelessly by the gentle breeze, pivot through the night like cells through the chamber of the self—the body's dance of atoms, chaperoned by foreign microbes, amoeba, viruses, fungi.

We're not individuals, we're colonies

It has taken a long time for biology and medicine to arrive at the idea that significant portions of an individual's own body are foreign to it. Now, however, microbiology in particular is discovering that there is no reposing, solid core within us, but rather a lurking void around which life's dance unfurls. In the human body, thousands of different players make the meaningful whole possible. We know that our body is colonized by microbes, particularly in the gut, which perform metabolic processes essential to our lives. Within our body, we carry our own, developed ecosystem, without which we could not break down and digest food. There is a reason that biologists call the "biofilm" of microorganisms that cover moist surfaces "bacterial lawns." With hundreds of species entangled on them—consuming, eliminating, extracting, and synthesizing matter— these bacterial lawns, like the Ligurian pastures, have the characteristic of an undulating meadow in the spring, inside of us. No wonder we have a feeling of recollection on such evenings.

In this age of advanced gene technology, the true abyss of renunciation from which we speak "I" is only now becoming obvious to us. For only a few years, it has been clear that bacteria are completely dominant in a healthy human being: On top of our ten billion body cells, there are one hundred billion microbial cells that play a role in our metabolism. This enormously increases the options for our bodily processes: If we include the microbes' genes, then we have over one hundred thousand genes at our disposal, as opposed to just over twenty thousand.[4] This sort of bacterial aid leads, for example, to children in Papua New Guinea being born with nitrogen-fixing bacteria (like those found in some plants and algae) in their intestinal tissues. This allows them to subsist for years on a plant-based diet without suffering from symptoms of deficiency.[5]

From this, the American microbiologist Bruce Birren concludes that "we're not individuals, we're colonies."[6] And these colonies develop sensitivities collectively: The type of bacterial ecosystem that lines the intestine will partly determine how successfully we absorb nutrients. Patients with a tendency toward obesity have particularly efficient bacteria. From a bite of cracker they are able to extract all of the nourishment that simply slides unabsorbed through the digestive tract of slender types. Even the balance of our neurotransmitters and hormones might not be controlled solely by our brains and bodies: "Could a person's happiness depend on his or her bugs? It's possible. Our existences are so incredibly intertwined," muses Birren.[7]

With these symbiotic microbes, our existence joins the ranks of a continuum shared by many other beings that exist outside our bodies. For bacteria are engaged in constant exchange with one another. During times of crisis, they share advantageous genes with one another like children sharing candies. This is why researchers nowadays are speaking less about the various types of bacteria in the world (as they are so transformative) and more about the diversity of their genes and the biological abilities they facilitate. Biologists are regularly stunned by this diversity: The US researcher Norman Pace investigated an ounce of silt from the hot springs of Yellowstone National Park in the 1990s and found more genetic diversity there than scientists had previously assumed to be present in the entire biosphere.[8]

This diversity is not neatly divided between distinct species or types but is available to all microbes within the context of symbiotic processes of exchange. The late biologist and symbiosis researcher Lynn Margulis believed, for example, that this exchange relationship meant that we should actually speak about all the bacteria on Earth as though they composed a single biological subject—one body swarming with countless cells. Consequently, we who are dominated by a bacterial ecosystem ten times larger than our own body's cells also belong to the great continuum of life. We are literally, physically, a part of the landscape. The moment we take sustenance from it, we enfold it and its inhabitants into our bodies.

For Varela, this pluralism and otherness at the center of the self was a lifelong puzzle—one that a good scientist did not simply push off the table, but rather one in which the principle of biological existence reveals itself. For him, a being is fathomless, a sort of spiral whose firm edges are delineated on different levels by various agents: the cells, the organs, the body. But in the middle of the vortex engendered by a life-form in the material world, there is a void. At our center, which comprises nothing, like the hollowness in the middle of a whirlwind, we fall back into the world. Every being is so deeply rooted in others that it is never identical with itself in the final analysis—its essence comprises far more what it is not.

This same nothingness delights us when the meadows lie quiet and full below the night sky, when the small glowing insects fall into their shadows and fade like dying stars. The meadow is a part of our body, folded outward, ready to be strolled through. It is one of our sensory organs in which we feel something that we would not otherwise understand properly, certainly not now, in this age of competition, of forced individualization, of lone warriors battling.

The astonishing thing is that as soon as we no longer determine biological laws using test tubes and electrophoresis benches, but using our sensory perception as living beings (which is perhaps more exact), every encounter with the world of other life-forms contains an unexpected lesson that far surpasses the findings of academic biology. Ecology that comes from the erotic contact of our body with other living beings discloses itself as a guide, tested and revised over millions of years, to how life is formed through the coexistence of many beings. Every ecosystem—both outside and "inside," in our body—vividly illustrates that this biological reality is not composed of blind, "deterministic" chains of commands, but results from the cooperation of a myriad of intractable actors, each of them following its own bliss, yet only to the extent that it does not damage the greater whole.

The life in which we participate so emphatically whenever we let the dark plants stroke our bodies is not a kindergarten where everyone is always pleasantly petted; but it is also not a merciless battlefield between unrelenting combatants, "red in tooth and claw"[9] and "at war, one organism with another."[10] The world of biology is more like a wild playing field with anarchic elements, where the rules of creative togetherness are constantly being renegotiated, where gang wars break out between little groups of co-conspirators and schemers, but also where one finds acts of magnanimous sharing, heroic dedication, and dreamlike bliss.

Four principles of attachment

In that evening meadow, in addition to the joy of seeing our own emotional metabolism wordlessly laid out before us, there are perhaps a few useful rules to be found that are also true of our coexistence. They could help us to improve our intercourse with other beings and to change the way that we budget the given resources. In other words, they could help us survive.

How might these rules read? They will continue to be a theme throughout this book. So right now I will offer a preliminary overview of them. The criteria for a first, foundational erotic was articulated in the previous chapter: Within matter itself, there is a manifestly inherent demand for deeper attachment, new details, a higher realization of its own imaginative potential. I therefore call this tendency:

1. The *Principle of Touch*. This describes how all bodies feel drawn to one another without intention or aim, how the drive to enter into relationship is dominant even within inorganic matter. At the same time, one can observe a countervailing tendency. I call this:

2. The *Principle of Freedom*. This means that as many individuals and species will come into being as possible, that the

whole will become ever more diverse, making it thereby more stable and stronger. Nourishing this whole requires a further principle. I call it:

3. The *Principle of the Gift*. This states that nothing in the rondelle of life is ever owned. All genes, for example, can be exchanged, free of copyright; energy streams as a gift from the sky and is distributed among all the levels of being that derive sustenance from it; and in the end, the death of an organism becomes a gift to those that feed on it. Everything essential is gifted, not that it might be hoarded for a monopoly, but that it might be shared as common property. Within the living realm, these gifts are given largely without design or intention. At the same time, we all depend on them. We need the gift, we need the sunlight that happens to fall upon us, the smiles of other people who like us whether or not we did anything to deserve it. Put simply, in the words of Michel de Montaigne: "because he was he, and I was I." The gift of the smile, given for no reason, invites us toward the other and guarantees:

4. The *Principle of Sharing*. This says that in biology, every "I" has been enabled by a "we." In such a system, every act of divestment amounts to an amputation. The "whole" is present as a constant and ineluctable part of the self. It is a part of its life processes—just as, conversely, thriving individuals are necessary for the establishment of a successfully integrated system.

Those who find all of this to be too much nocturnal speculating can gladly return during the day and walk up into the meadows, equipped with spades, specimen cups, and reagents. Good biologists do not allow themselves to be misled by appearances—they dig deeper. They do not stop at being poets; they are researchers. They clear the vegetation and analyze the ground. What reveals itself? The scientist's hand rummages through a space comprising not just minerals, but also crumbs from the organic realm. For up to one-third of the ground can be composed of organic substance.[11] It is made up of humus, bits of leaf, little worms, and crawling insects, streaked with a myriad of ever-finer roots, along with invisible mycelia of fungus. Everything is intertwined and forms a single, impenetrable body, a bottomless net, a feeling skin spread over stone.

Gregory Bateson was fascinated by the fact that the relational networks between root hairs and mycelial filaments, between predator and prey, partners and competitors, have a form similar to the neural pathways between the different hubs of our brains. Bateson drew several conclusions from this: that the landscape is also capable of thinking—not in ideas and words, but in forms, colors, tones, and scents. Its thinking has no object, and it therefore knows nothing of either accusations or reproaches. The natural world thinks by transforming itself as a subject. The relationships within an ecosystem thereby constitute something like the synapses of a landscape's nervous system (a very specific nervous system, which has the form of a very specific landscape). In this, an ecosystem resembles a brain. Like a brain, it is capable of cognition. The way in which vegetation changes as the climate around it becomes more dry, for example, could be imagined as the way in which that ecosystem *imagines* a drought. The biosphere is a system that constantly produces new relationships by responding to existing ones. Our brain does the very same thing. Moreover, since it resides within a body, it does not just map the relationships from the outside, but is itself a part of the relational network within an ecosystem.

Whenever neurobiologists observe that the brain is constantly learning,[12] this therefore means that for as long as we are alive, we are part of a process of mental and bodily growth wherein we interpret encounters and transform ourselves into the history of these encounters. The brain is thereby a reflective organ of the world, comprising primarily relationships. It reflects these relationships by producing relationships *within itself*, by establishing relationships *to the* relationships in the world, and by attaching new relationships *onto these* existing relationships. The brain is an organ that reflects the world by simultaneously making itself into a part of this world.

And all of our thinking, for its part, forms its own ecosystem as well. Mind is an ecological phenomenon, the result of a collective dance. The open country dreams the forest in the form of a hedgerow, as we saw in the last chapter. The meadow dreams the insects by choreographing a ballet and reacting in it to alterations in its surroundings—changing itself and making these changes visible. When we roam through the rustling grasses in the dark, through the quiet meteor shower of fireflies, we are dancing with them.

The irresistible longing to be

A meadow, dancing this slow dance of infinite possibilities in its blooming summer dress, is the erotic locale *par excellence*. Woe be to all the hayfever-plagued inhabitants of urban modernity who cannot indulge the unbridled impulse to kiss another person in such a meadow, obeying the bidding of the balmy night and the titillating blossoms. The meadow embodies what the British author Aldous Huxley describes as our lifelong longing: It is full of gracefulness. Huxley believed that we, and all other creatures, are seeking "grace"—a word that means both loveliness and blessing.

The longing that Huxley speaks of is not a metaphysical fiction; on the contrary, it is integrated into the biology of organisms. Of course, to understand this, we have to get used to looking at life-forms with new eyes—eyes that no longer consider them as comprehensively optimized mechanical artifacts, but as beings that want to survive, beings for whom a continuing existence has an absolute value that suffuses everything that they do with feeling: with a desire for being and with a fear of failure.

In my book *The Biology of Wonder*, an essay on the subjective feelings that pervade the world of organisms, I described this foundational emotionality in "Three Laws of Desire." The laws of desire frame the principles according to which life-forms experience all instances of bodily concern as existentially meaningful. A life-form can fail at any time—and therefore it *wants* to survive. Because of this existential life wish, the world of organisms is not a neutral stage, but one deeply steeped in values and meaning. Its principles—the *wish for continued existence*, the *visibility of this life wish as an emotional expression*, and the *necessary presence of other beings to enable one's own life*—make up the ground rules of desire according to which living matter pushes toward unfolding. These rules are true equally for processes of material exchange and for mental and emotional experiences.[13]

This desirous longing unfurls out of the yawning abyss that opens between the aims of an organism striving for wholeness and the indifferent matter from which it must repeatedly assemble itself. In *The Biology of Wonder*, I attempted to describe the organism as something more than a machine made of matter. Instead, I tried to understand it as a complexly assembled piece of matter's particular striving to not only maintain its existence, but also to expand. The life wish is not a program, but an urge that emerges out of matter and also structures it. A being—even the simplest cell—*is* this longing.

Therefore, an organism is always already something internal, something nonmaterial. It is nonmaterial only because it is composed of pieces of matter that, as we know, have their own independent existences. The complex substances that make up a cell would, in the blink of an eye, collapse into simpler, "dead" parts, were it not for the presence of a constant, structuring order that strives for continued existence and holds together the independent components. This coherence is also not an achievement of the "gene," for the gene itself depends on cellular metabolism in order to be maintained and repaired.

The inmost core of an organism is thereby composed of a striving to maintain its own wholeness, rather than disintegrating into individual fragments of matter. We all know what it means to feel this striving, because we, too, want to go on living—and not only to live, but to unfold ourselves in a productive way. We know the life wish from within. As soon as a being—even if it is just one cell—produces itself as a whole out of the same matter that could, at any time, consume it once again, this longing exists in the world: The irresistible desire to be, the desire for more life, for healing, for plenitude, for wholeness. For that reason, no animal—and also none of the simpler life-forms that precede it—is a machine. It is instead a self-interested self, caring for itself, a self for whom everything has meaning because everything bursts into the sphere of its life as something positive or negative. Every being therefore reiterates the basic configuration of feeling subjectivity. Every being is one of subjectivity's masks, it is a form of being a subject—a feeling, longing, vulnerable, triumphant subject.

This omnipresent subjecthood confers onto the natural world the characteristics of a psyche. The emotionality of the natural world is not a metaphor. It is not a projection from the symbolic repertoire of our cultural history, nor an arbitrary thought, but the inside, the existential side of the natural world's biological functioning. It is its erotic. Simply put, it is the vibrating quality of living that materializes as an individual, as a self for which every encounter in the world has meaning and appears as though colored by feeling, not as "neutral information."

Feeling thus becomes the physics of the organic world. As life-forms among the other beings of the natural world, we are ever observers of an aliveness that we also carry within ourselves. We see the outside of what we experience as value, meaning, and feeling "inside"—in our souls. We see this outside and know it to be an appearance occasioned by an inside, and that we can perceive this inside with our senses. The natural world is consequently more than just a reflection of soul. Filled to the brim with life and striving, it is a psyche whose outside gives us access to our inside. Every being is a figure of our inside; no: Our inside is one of its possible expressions. We—feeling, analogical, formative beings—can only understand this inside by seeing it as a living image of ourselves before us: as a happiness or a sorrow, which likewise fills us with happiness or sorrow.

Seeing in plants and animals the outside of an inwardly felt aliveness is the only way we experience what it means to be alive. We are hidden from ourselves: Our body, which enables us to proceed with feeling into life, simultaneously keeps us from encountering life as something external. We can only understand by participating. It is like the blind spot on the retina: We are not able to see at the very spot where the optic nerve, which transfers the images to the brain, connects. We need the feeling embodied in the body of another being in order to make full use of our own feelings. Just as we recognize many emotions for the first time in the beauty of a work of art, contact with other beings plumbs the depths of our hearts and makes us fully ourselves.

An endless thread, knotted onto itself

The erotic of sharing is thus not only a longing for nearness. It is also the shock caused by this nearness—a shock at how it prompts the response of an other, in whose eyes I read: I am at home with you. The erotic of sharing causes my field of vision to expand. This is the outcome of all forms of attachment—we will investigate this in the second part of this book. It is actually the prerequisite for our own identity. This identity—that of a feeling, biological life-form—begins with other bodies that mean something to us and thereby become components of our own experience.

We human beings see with plants and animals just as poets see with words. Deep down, other life-forms are what make our own perception possible. When it comes to the realm of our own identity, we are blind, deaf, and dumb without other creatures—robbed of sensory organs important to the act of being. Without them, we *are* not, for all being is reciprocity and reflection. Only *with* others are we alive, present in the enlivened flesh of this Earth, which brought us forth and carries us. We have too quickly forgotten this fact. Now that many of us procures its food processed and packaged in the supermarket, it is no longer obvious that *we nourish ourselves with life in order to be alive ourselves*—physiologically, as well as emotionally.

So the beginning of spring reposes not in my imagination, nor in the rosy glow off the granite, nor in the mistle thrush's mating call, but rather in all of us as we collectively surrender ourselves to it. The call of the birds, of the blackbird and the thrushes, make up a placeless choir that invokes something in the here and now that is not there, and perhaps never will be. They are voices, physical oscillations of the air, and at the same time—pure imagination. They are inside as well as outside—both, and neither alone. The psyche that lies before me there in the natural world, that reveals itself as the outside of an inside, is only there before me because it is also in me. We realize and specify one another mutually—the perceiver and the perceived, which also perceives me as I experience it.

Life is something that brings the inside of a body to expression: the inside of the world, ever present, the potentiality of matter that dreams of awakening, regardless of whether this awakening is accompanied by pleasure or pain. The present tense of this inside is poetry. We don't know what it will bring into being in the next moment. It waits in the little sonic bursts that make up the murmuring of the gray and green Vara river, glossily flowing over rose-hued and delicately lined black stones. Poetry is the whole that we recognize, that we know as ourselves, even though we could not pound out this knowledge into formulae: The sole possibility for understanding here is to answer in the form of a gesture, a look, a bodily posture—to answer with creation, with poetry, which is nothing other than the whole, which is nothing other than a stray fragment.

The slivers of the mistle thrush's song are reflected in the glistening water of the Vara river, in the crushed stone of its granite pebbles, in the echo of the mountains, suffused with a wild ruddiness for the first time today. The carbon, the silicon, the water, and the air interweave to form a net without beginning or end. They form a single, endless thread, knotted onto itself.

— *chapter three* —

DEATH

*The audacious thought of the necessary imperfection
of all creation.*

GERSHOM SCHOLEM[1]

$\mathcal{E}$ ven at half past ten, the sun was still blazing. The evening seemed as though it did not want to end. We had all gone into the smoke sauna. It was part of the ancient farmstead of Leigo, a little building of rough-hewn logs with a thick mossy roof beneath a few tall linden trees. We—that is, the somewhat younger philosophers and biologists participating in a conference on an alternative view of life. All those who, in this endless light, felt a desire for adventure—for a little, civilized adventure—had disrobed beneath the trees. It was clear in the slanting sunbeams that the night would be filled with light for a while yet, that rather than give way to darkness, the light would hardly recede at all, that instead things would simply become more transparent, more translucent.

In front of the blackened log cabin, where heat hung glimmering in the air before the door whenever someone entered, lay one of those glassy, dark ponds so common in central Estonia. The little body of water seemed to be a creative space in which nothing was as yet decided, but whose surface seemed to be waiting for concentric rings of gently dimpled water to skim across it like fleeting ideas, thereby stirring the "reservoir of darkness," as the British poet W. H. Auden once wrote.[2]

The world manifested as potential. It was cooling and soft for our sweltering bodies as we periodically bolted out of the smoky heat of the sauna and over the soggy embankment to plunge into the dark water. A clever someone had deposited a few cans of beer in the weeds on the embankment. Then we sat again in the dark hut, sweating on the wooden benches blackened by centuries of smoke. Two female researchers from Italy took part in this exaltation in bikinis, but ultimately stayed outside the sauna after one of the Estonian scientists told them in a soft voice that the air temperature in the sauna would instantly cause their bikinis to shrink tight against their bodies.

Once the alternation between the two extremes had exhausted me, I sat with my beer in the diaphanous evening, my back against the logs of the hut. My colleague, the biosemiotician Kalevi Kull, sat down beside me. With his unique way of turning incidental details into subjects of philosophical inquiry, he began to question me intensively about why the oscillation between hot and cold—or rather, between deadly hot and dangerously cold—brought us so much enjoyment and bodily pleasure. He wanted to know what I felt, when I felt it, in detail.

Kalevi, who teaches at the University of Tartu in Estonia, is a Socrates of biological self-awareness. If you start to answer this scholar's questions you can be certain that at some point you will end up entangled in self-contradictions—and precisely at that moment you will grasp something new. As it happens, Kalevi has a wonderful description for this type of conversation, which for him accomplishes everything science can ever hope to: "Helping one another in mutual understanding," he calls it.

Kalevi is the heir to a long tradition of biological thought deeply entwined with this country on the coast of the Baltic Sea. To this day, in this little nation on the border with Russia, where birch and pinewood forests harden with jingling frost during the long winter months and where the summers are lit by a never-expiring light that illuminates the sparsely populated landscape at night as though it were glowing from within, a tradition continues of seeing life-forms not as efficient machines engaged in genetic competition, but as "thoughts of nature." This is how the evolutionary biologist Karl Ernst von Baer described it during the nineteenth century, at any rate. The most important heir to this thought was the biologist and philosopher Jakob von Uexküll, who taught first in Tartu and later in Heidelberg and Hamburg. He is the great uncle to the younger Jakob von Uexküll, who lives in Stockholm. This younger Jakob founded the Alternative Nobel Prize in the 1980s and awards it annually to people unswayed by a worldview that only rewards that which is expedient.

No light without darkness

The reason Kalevi prodded me so much about the opposition of the two elements of our chaste bacchanalian enjoyment that evening had something to do with the questions that he was interested in that summer. How much opposition can a life-form tolerate within itself? How much opposition is necessary for the processes of life to progress at all? How crucial is this idea of opposition, or paradox, to our understanding of life? Kalevi goes one step farther: "A cell only functions," he said to me on that evening between a few sips of beer, "because it is incompatible with itself. Because its component parts are irreconcilable. Every cell is its own contradiction as long as it is alive."

"So, for the individual parts of a cell, does living mean having to help one another understand each other?" I asked in response.

"Precisely." He reflected on that for a while and looked into the distance with a whimsical expression. I didn't know whether Kalevi was thinking or whether he just did that to keep me from being overly embarrassed by the fact that he already knew everything that occurred to me during our dialogue. "In a cell, dimensions that are fully irreconcilable come into contact with one another. The genetic data—which is an abstract code, and the cell body—which is a concrete, material being in space. Both are incompatible with one another. And this incompatibility means that one always has to be translated into the other."

"Then there is always an excess that cannot be transferred," I thought aloud. "Understanding fails. But in the logic of this antithesis, it only has the chance to be 'understanding' because it must ultimately fail. Were it not for the ever-present possibility of death, beings would not need to be possessed by the urge to evolve, to go on existing. Without death, the being is a machine. Or more generally: Aliveness must be able to fail if it is to be truly alive. Only because of death does life become creative."

"Yes!" Kalevi cried, slapping his knee. I knew that he had thought of all this long ago.

In the lavish summer sunlight that spread between shadowy trees and seemed to dissolve the transparent blades of grass into the surrounding air, my own ruthless assessment seemed unreal to me. Was this very moment not one in which everything seemed to be in harmony? In which *harmony* actually proved to be the true character of the world? But I knew that Kalevi's surprising hunches were usually accurate.

And perhaps that evening at the height of midsummer was, in truth, an example of just that: of the necessity of death that would cause the pond to freeze over and stiffen the linden trunks, so that after a while they might capture our hearts once more with the ecstatic hopes of spring.

In my mind's eye, I saw the area around the sauna hut in winter, dusted with snow, a frozen silhouette in black and white. It was almost inconceivable. And yet, unavoidable. The northern summer's splendor owes everything to the long, gloomy winter. "All light must be informed by darkness and all success by suffering," the philosopher and mystic Richard Rohr observes.[3] This attitude—the ever-present belief that you can rescue yourself in the security of just one side, ironing out painful contradictions once and for all—robs us of our aliveness.

The English philosopher Alan Watts says: "By and large Western civilization is a celebration of the illusion that good may exist without evil, light without darkness, and pleasure without pain, and this is true of both its Christian and secular technological phases."[4] This would mean that wherever there is beauty and productivity, there is also is also a bleak, abysmal side that we cannot avoid. Everything else is an illusion. And it would also mean that after two hundred years of intensive attempts to bring about "enlightenment" and "illumination," now that our civilization has put the Earth in a position more unsettling than any it has been in for the past two hundred million years, perhaps the moment has come to say goodbye to our belief in a life without death. But this farewell, I thought apprehensively, will be itself a death.

In the days following this casual but also momentous summertime conversation, I went looking for examples. I tried to find them not in mysticism, nor in my own terribly limited and often unexamined experience, but in the science of the living. Those days, I worked on an empiricism of the incompatible things within us, on a biology of death, an ecology of contradictions. In short, I sought to understand whether Kalevi's statement, which seemed ludicrous at first, could possibly be true.

If it were true, that would open new doors. The world of healthy life would then be a good deal more problematic than it had once appeared to be, but also a good deal closer to what it feels like on the inside, in my experience: a challenge to turn contradictions into a narrative that produced meaning.

I tried to research how much death shaped life and the extent to which the most productive life included death. In the process, I had to think a lot about that little poem by Rilke, "Closing Piece," which had always moved me when I was growing up, even though I was not sure I understood it: "Death is great. / We are his / with laughing mouth. / When we think ourselves in the midst of life, / he dares to weep / in the midst of us."[5]

An ecology of death

Slowly, it became clear to me that the negation that Kalevi had discovered in the heart of the organism was not the only one. I discovered that life, in truth, is a whole network of such incompatibilities, and I began to get used to the thought that this is perhaps the only reason that it functions at all. For life is, of course, just dead matter that has become suddenly obsessed with actively reproducing and synthesizing itself into ever more improbable forms.

The great incompatibility at the heart of our existence begins long before the incompatibility of genetic data, with the minute bodies of each cell into which it must be translated. The deepest incompatibility is to be found between the autonomy that every life-form, even the simplest bacterial cell, introduces into the world, and matter that is organized according to the laws of cause and effect. The real scandal is that these willful cells exist at all, and that new ones are being created every moment out of a substance that would gladly have rested on the ground as lifeless dust for the next billion years. And perhaps this division extends even into the deep structure of matter, forming a crack that divides substance from the force of inertia, which constantly strives for the lowest possible level of energy and is therefore caught in a rivalry with the secret desire for unfolding and resonance.

The self-organization that I spoke of in the last chapter, as well as its breathtaking results—from the perfectly balanced uranium nucleus to the slowly undulating snakelike arms of feather stars on the ocean floor—are, it seems to me, direct echoes of death. Form arises because the forces that want to raze everything to the ground leave few outlets by which creative identity can assert itself—only thin lacuna and narrow canals in which the creative desire coagulates, filling them as liquid metal might fill the veins of an ancient anatomical preparation that has long since decayed, leaving only the metallic network behind. Things' desire for complexity is a reflection of the end that must someday consume them. The facticity of forms, so profound as to bring one to tears—the fact that every leaf is different, every bud a unique individual—is both a reflection of the death that awaits all things and a witness to wholly unique ways of avoiding decay and celebrating the triumph over it. Both the wealth of imagination and the austerity of limitation belong together like the tides and the moon, like the woodland and the meadow, which must cocreate the hedgerows' whirling bliss of blossoms. The explosion of an inconceivable number of individuals is also the way in which life thinks death: Creative diversity is the perception of finite nature and its transformation into a gesture that mocks all thoughts of finitude.

Death alone gifts to all existence its indivisible uniqueness. Death is individuality inverted. Hannah Arendt, the great philosopher of the human as a being of flesh and blood that can only exist through relationships, observed this. Arendt used the term "natality" to name the singularity of all creatures that move through stages of nonbeing, emergence, and disappearance. All existence represents a new beginning, a new answer to the insatiable desire for being; in every unique thing, this desire is not exactly fulfilled, but manifested in the boundless intensity of an urge to exist as a bounded material thing that will someday end. Natality ensures that all of existence takes on a deeply poetic gesture despite—because of—death's demand for nonbeing. It transforms the law of nonbeing into one of being-regardless, which wastes no words, but simply is. As carefree as the water ouzel, whose delicate and dogged life John Muir dreamt of following from beginning to end. As carefree as the wren, itself a featherweight, an almost massless ball of warmth that hurries along the icy posts of a ramshackle fence in the dead of winter, feeding on tiny insects.

In the history of spiritual philosophy, this singular being, this very specific "suchness," plays a central role. In the Middle Ages, the philosopher Duns Scotus taught that every creature could reveal God simply by being wholly and uniquely itself. To the eyes of those who look upon it with love and engage fully with its unique being, each creature releases the divine light awarded to it.

The concept that Scotus used for this phenomenon was the Latinate neologism "haecceitas," or "thisness"—which the British nature writer Robert MacFarlane also translates as "self-ablazedness." Thisness manifests the creative triumph over nothingness, a triumph that must, however, be perpetually wrested from that nothingness. One truth is manifested more in the simple being of bodies than anywhere else: that a force pervades the world that craves existence, and with it individuation and self-awareness. The philosopher Gottfried Wilhelm Leibniz observed that philosophy begins with our astonishment at this—the question, why does something exist, rather than nothing.

Those who, like the mystics of the Middle Ages, quiver with faith in the goodness of things have no need to deliberate before arriving at an answer. They must only await the end of winter, for example, when the flowers of spring snowflakes appear beneath the linden trees in the swelling meadows, crowning the thick green with a swarm of white blossoms. The way they emerge from the ground—unexpected and completely unbidden, standing there full of naiveté—is an experience of pure grace in whose presence the splendor of creation unfolds with ever greater complexity in ever more delicate forms. Upon closer inspection, the sky is folded into the calyx of every petal. This is it, the overwhelming experience of simultaneity, the synchronous presence of all possibilities—of growth, maturation, and decay—the totality of being, the newborn laughing coyly in the face of disintegration with the causal ease of a child's tears, all of this concentrated in the damask of a single petal.

Seen in this way, every life-form becomes the center of the universe, a singularity in which the essence of creation is fully revealed: namely, the desire to be that takes form as desire, not as sovereignty or material self-mastery. This is why life is easy to destroy. Within itself, life is already carrying death, which is just waiting for the door to open to it; life is a delicate web of desire for a completely purposeless, unsanctioned uniqueness. But because this desire directs every stage through which substance transforms itself, through which the atoms shift position, it becomes the indestructible mold, the source of all forms. Goodness prevails, so to speak, but it prevails at such a low level that we must have a good deal of patience with it. Learning this patience would mean learning to love; it would mean taking life by the hand and making ourselves into instruments of this profoundly healing aspiration.

At the end of the nineteenth century, the Irish romantic poet Gerard Manley Hopkins centered his work around his astonishment at the world's endlessly repeated, yet always varied thisness. His language trembles with amazement at this constantly incredible gesture of revelation, and in the end, the only response suited to this incomprehensible phenomenon of a desire that abides as desire is to echo it. Language becomes an instrument not of analysis, but of serene acceptance. As Hopkins writes in one of his most famous poems:

> *As kingfishers catch fire, dragonflies draw flame;*
> *As tumbled over rim in roundy wells*
> *Stones ring; like each tucked string tells, each hung bell's*
> *Bow swung finds tongue to fling out broad its name;*
> *Each mortal thing does one thing and the same:*
> *Deals out that being indoors each one dwells;*
> *Selves—goes itself; myself it speaks and spells,*
> *Crying* What I do is me: for that I came. [6]

Farewells and fresh starts:
our stuff changes, our self remains

In order for desire to be able to unfold in a body as a creative, deeply vulnerable gesture, it must already carry its negation within itself. It must be an urge that knows of its own impending end and is thereby free to pursue the hazards of growth and attachment, and to risk itself in the process. To speak with the words of Kalevi Kull, desire must be incompatible with itself, internally broken, an impossibility—and at that precise moment, it can assert itself so intensively that we perceive through it, so to speak, the creative urge of the entire cosmos: the burning scarlet dragonfly, the kingfisher of gleaming metallic blue and green, which is a flash of lightning and a fleeting thought and yet nothing more than a fragile body, an aspiration of matter to be more than simple substance, lasting just five short years. And if we shift our gaze from Gerard Manley Hopkins's ecstasy to the laboratories of biological science, we can see also with the cool view of a researcher that precisely this—this rupture—constitutes the life-form's biological character.

Hans Jonas, the Jewish biophilosopher (and close friend of Hannah Arendt) who left Germany for good before World War II, vividly described the improbable risk that matter undertakes when it follows the desire for more being.[7] Jonas does not let go of a detail that biology has completely overlooked, intoxicated by its success at describing all beings as "genetic machines," and climbing to the top to become the leading science of the twenty-first century. This detail is *metabolism*. This word captures the curious phenomenon that the matter composing every cell (as well as our body) is constantly being rejected and repelled. While in the prime of life, every cell is dying a continual death: Its matter, the molecules that it used to build itself, must be destroyed and released in the next moment so that life's flame is not extinguished. The cell, as it is, in its material, must die in order to live. The term *metabolism*, originally derived from the Greek *metabolē*, describes this revolution. The direct translation would be "throwing over." In order for a life-form to be able to experience something, everything must be overthrown—in its body and in what it ingests.

Do you still remember your days in school when the biology teacher drew the so-called citric acid cycle on the board? This circle of biochemical reactions details the central "energy motor" of every cell. Nourishment—a little bit of sugar—is introduced into the cell, energy is released from it, and carbon dioxide is emitted. That's it for the boring lesson, the learning objectives of classroom work. Of course, the teacher withheld the most important part (because his professors had also failed to show it to him). The emitted and exhaled carbon dioxide is not the exhaust of combusted "fuel," as in a motor. The carbon atom does not come from the ingested nourishment. It comes from somewhere else, from the cell itself, from the cell's own body. So metabolism means: I subsist on what becomes my body, and I exhale into the air what was my body. I am the grain of the field that died for me, and I die constantly and transform myself into what the plants inhale, such that my body becomes their new bodies. The organism is a closed being, and at the same time matter flows through it. Matter drifts through the bodies of a vast array of organisms without ever being identical with them. A carbon atom in the calm grasses of the meadow was once a part of the air, and before that an insect, fruit, perhaps a human body, its breath, perhaps me.

Isn't that a genuinely erotic relationship? A bond that produces deep inwardness—and that demands, at the same time, complete self-sacrifice? The functioning of the circle of life on Earth depends solely on the fact that we all share in the great body of matter and pass through one another reciprocally. Life is touch in a much deeper sense than just touching skin to skin, colliding against foreign masses: It is touch as penetration of one by another. The existence of each one of us—plants, animal cells, I as a human being—depends solely on the mutual related-ness manifested in this exchange.

And this exchange only becomes possible if each body does not remain materially the same (like a bronze statue or a diesel motor does, for example). In contrast to an object or a machine, a body regularly splits off a part of itself in order to survive and incorporates a piece of the foreign world into itself. This is precisely why it is wrong to compare a life-form with a machine: A machine does not metabolize. The fuel that I put into the tank burns but does not transform itself into another body. The carbon atoms in the gasoline are the same carbon atoms in the exhaust. This is why combustion engines warm the atmosphere with increasing levels of CO_2, while living, edible bodies in an ecosystem do not.

Metabolism entails ongoing striving and birth. Who we are, therefore, is not determined by our material substance. That is the sublime knowledge that Hans Jonas imparts to us: Who we are is defined by our individual desire to become, to grow, to unfold, and to declare ourselves. Life is a process in which an identity generates itself. But this identity resides not in matter, but in the desire with which it births itself in each new moment, in whose name it undertakes every attempt to unfold itself creatively. What composes us is a gesture, an act, a wish: the wish to continue existing.

And this wish is precisely the opposite of matter, which is always striving for its final resting place. The desire for existence, which we can only realize through a substance that is continually seeking to escape this desire, is the deepest contradiction of the living. This desire likewise describes the structural break that will someday lead inevitably to catastrophe, to the death that awaits all of us in some form or other.

If we consider exactly what makes life-forms "tick," it becomes clear that the strict differentiation between outside and inside, which our civilization holds in such high regard, simply does not exist; rather, matter organizes itself complexly. And complexly organized matter inevitably creates a sphere of meaning, that desire for more being that characterizes the space of subjective experience. Consequently, outside—pure materiality—and inside—the experience of good and bad, feeling, being affected, the striving for aspirations (in other words, everything that characterizes our condition of being)—are by no means different spheres or distinct worlds.

Outside and inside are two sides of the same thing and indivisibly bound together. Not because "everything is spirit." Or even because there is no "true" death. The whole is not at all esoteric; it is fleshly. Esotericism is after all just another altogether human refusal to accept death by taking control over everything—in this case, by means of the illusion of a "spiritual" ability to see through and master the whole.

Only *because* a life-form must one day founder does it transform the outside world into an inner world. Life-forms are bodies among other bodies that "want" to maintain themselves as they are—as bodies with a particular constitution, in good health, and with good prospects for the future. (Single cells and primitive organisms probably do not "want" this in the same way that we want things, but they behave according to an urge for survival that we can very easily relate to.) Something that has at least the urge to survive and consequently fends off tenacious disruptions is, by definition, a subject, and not a machine. It is a being that pursues a value. Everything that supports its goal has a positive significance in the organism's experienced world; and everything that hinders this goal, a negative one. External touch thereby becomes—in an admittedly rudimentary fashion for simple beings or single cells—an existential experience of good or bad. We can therefore say: Beings that have the goal to exist develop needs. We call the experience of having these needs and seeing them fulfilled to varying degrees—*feeling*.

Desire opens the window that makes experience possible. Only a being that can fail experiences the existential significance of what it encounters. A statue knows no meaning, because it is not mortal. For a desiring subject, every crackle of reality contains a message, an answer to the anxious question of whether the next moment will simplify or hinder existence. Death is unavoidable, but only death makes the world legible. And this language, written in desiring, breakable bodies, is understood by all beings equally. The death that awaits us all is the only thing that makes this world into a collective affair. Metabolism transforms matter within our collective body. And its vulnerability forges us together into a collective spirit.

Considered from the standpoint of biology, this is the sympathia universalis that the Renaissance scholar Fracastoro saw permeating the universe—its omnipresent erotic of living in flesh and blood. Its Eros is that urge toward plenitude and integrity that each being combines with matter, that desire to unfold all of one's own particularity and thereby become *self*—a unique birth that asserts itself in inimitable fashion against the regression into the stillness of death.

This is the spark of Eros that keeps reality alive and that glitters back at us in all that lives and all that enlivens us: the ancient urge, freshly born in childish grace, to be oneself, to express oneself, and to feel the whole pulsing through one's veins. The erotic of ecological connectedness can actually be felt. It is what a being senses first and foremost when it opens its eyes to this world of creative desire and imminent demise. It is the life impulse that makes every heart beat more rapidly.

If we want to love, we must learn to carry forth this Eros. We must give a part of our own ecstatic individuation back to the world, must therefore be prepared to die, to some extent, so that something else might be—for example, the landscape that nourishes us, or another person, like our own child or our partner.

The tragedy of the organic

The relationship between life and death is not simple or unambiguous. Stubbornly insisting on life can result in the opposite. The frantic desire to ward off death can actually invite it. Conversely, if you wish for life you must be prepared to welcome death. If you demand life, you must accept that death is a part of life, its dark half, without which no life, no experience, no living meaning, no poetry, and no love is possible.

This viewpoint is no myth. It is gained through biology, by looking at and observing the simple ways in which a single cell stretches and extends and moves beyond itself. This viewpoint attempts to stop fleeing from reality and to radically accept it—as radically as the inexhaustible water ouzel that dives from the icy river stones to seek nourishment, as radically as the mistle thrush singing timidly on that first early spring day in the Ligurian mountains, which surrendered themselves to the evening light and the storm surges of spring, just as they gingerly lay themselves down in the lap of the harsh winter and its blizzards.

The perspective on life revealed here is tragic, not optimistic. Archaic, not technocratic. It is not just sunny, it is harshly bright. It is harsh because it internalizes the fact that there can be no creation that is not bought with defenselessness. Perhaps we can speak of a biocentric tragedy, of a rift that cleaves the living heart and makes it bleed. But this division alone is what makes space for desire, it frees this desire, this longing to reveal itself in snowstorms of hawthorn blossoms, in the swifts' joyous arcs through the evening air, in the newborn's smile as it receives for the first time the gift of this world and all of its unfathomability.

All of us—all life-forms in this biosphere, and indeed all things in this reality—have death in common. It is a profoundly connective element between us. It alone allows us to understand the trembling of another being's tormented limbs, even if that being is as distant from us as a tadpole: The jerks of its small, smooth body in a nearly dried-out puddle let us feel, in our own neural pathways, what it means to be no more. And the blissful grunts of my little poodle, turned on his back so that I will rub his warm stomach, likewise teach me to appreciate that for this animal and all other beings, the answer to being-no-more exists—as it does for us—in the smiling triumph of joy.

In death we are all equal. We are all siblings in the face of it. It is nonsense therefore to claim that an animal does not suffer or that it is not witnessing its own suffering. The world of other beings is an open book, because we are all gripped by the same *conditio vitae*, must all exist under the same conditions of aliveness—and of death.

Seared by the light

It was May in the mountains of the Ligurian countryside. The evening of a sunny day began to coat the mountains with a veil of fog, behind which the darkness steadily advanced. Olive trees and sclerophyll bushes, which keep their leaves the whole year through, do not grow here in the Apennines as they do on the coast a few kilometers to the south. Here, oaks, sweet chestnuts, ashes, willows, and cherries cling to the rugged hillsides. In the winter, it is a gray world, a world of colorless pastel. But in spring, the dance of newly shining light, of the fresh chlorophyll and the snowflakes of blooming umbels is all the more thrilling.

And just now the mountains have clouded over with a tender skin of the freshest green, with that green that is hardly substance, but almost purely light, as though it intended to categorically prove that everything the sun strikes transforms into effervescent brilliance. This same green allows us to inwardly follow how plants perceive the sun with the entire surface of their bodies by growing toward it. This light makes it comprehensible for us that photosynthesis provides us with life in a way that no theoretical explanation can. Its delicate illumination is the light that stretches irresistibly upward. Light pours out of the trees as the leaves open in the ecstasy of a few days, or even a few hours.

The ambiance reminds me of the desire that Theodor W. Adorno once described as he nostalgically opined that no urgent appeals to the neutrality of the scientific observer could "dismiss those cloudless days of southern lands that seem to be waiting to be noticed. As they draw to a close with the same radiance and peacefulness with which they began, they emanate that everything is not lost, that things may yet turn out fine."[8] It was the conclusion of a day on which I would have thoughtlessly declared that life always wins, that this biosphere is above all a realm of life and of living happiness.

Down below, the river's gray water, on its way to the sea, tussled with finely veined gray, blue, green, pink, and black blocks of stone. I heard it rushing. I walked through the sweet chestnut wood of immaterial green, gazing toward the meadows where, at this time of year, the cows seemed to pluck more orchids than fresh grass, gazing at that greenest of all greens.

A tingling sensation caused me to look at the back of my hand. Crawling over my skin, green and transparent, shining in the light like all of the tender leaves, was a little katydid. Eagerly, it ground its anterior legs against microscopic mandibles. Its bright green limbs appeared so delicate that my eyes could barely make them out—as though they were composed of a single line of delicate cells. The two black points in its eyes moved in sync with the long antennae as I nudged the animal with my finger and it turned toward me.

A katydid in the first stage of its life had somehow landed on my hand, still wingless and completely transparent. Through the animal's spring-leaf coloring I could see its delicate frame, as though its anatomical structure, too, was nothing more than an apparition of the light, a prismatic fanning of light waves out into undreamt-of details. It seemed as though the light itself, whenever it encountered a living will, was ineluctably inspired into such arabesques.

The katydid is one of those countless beings that hatch from a cluster of eggs well buried in the ground and suddenly repopulate the newly rewarmed world, as abruptly as if they had fallen from the sky, as suddenly as if they had been distilled from nothing but air and the desire for new life. From the end of July onward, after the birds have long since fallen silent, the adult specimen of this insect, having undergone several molts and now in possession of large, rustling wings, would sing its monotone song to the approach of the Italian autumn in an inexhaustible trance of parting, a whirring requiem, the patient accompaniment to an extravagant disappearance.

I considered the katydid at length and carefully. I looked at its gossamer, transparent wing sheaths. I saw the moving antennae and the green carapace, a shimmering surface, as completely present and material in the here and now as anything, and yet simultaneously a porous window, an entrance, an invitation into the inner workings of things. The animal made me pause as though thunderstruck. As though pierced by happiness. As the echo of the setting sun streaked the mountains and the trees, the insect offered a form of light that you could touch with your hands and yet never grasp, concentrated here and dispersed everywhere. Before me and within me. The insect was itself alone—and it was wholly me.

I carefully shook my hand after I had looked at the animal, let the being with its groping legs and quivering antennae float into the grass at my feet, that it might nibble with its tiny mandibles the green leaf tips and thereby nourish itself further with light. I straightened up from squatting and walked along the narrow path another hundred meters over crookedly trodden stones as the sun began finally to decline and its slanting light turned the greenery, glazed with cool mist, from yellow-green to orange.

On the way back, I casually looked down at the spot where I had entrusted the insect to the ground. A fidgeting on the forest floor, seen out of the corner of my eye, caught my attention. And there it was still, that little katydid filly. But everything was changed. The animal was having a wild fit, its body twisted and contorted. Convulsively, it jerked its hair-thin legs. For a breathless moment, I did not want to believe my eyes. Then I grasped its terrible plight.

The baby foal of a katydid was caught in the pincers of a large black wood ant. One of its jumping legs was already severed, the katydid's abdomen squashed, the being still living, but doomed for certain. The singing summer was over, the glowing green had passed, the colors extinguished, God never born. In the ever-weakening convulsions of the shining katydid, the light seemed to trickle out of the world; although it was still daytime, every color seemed to be soaked in black. It was something that I had never perceived before. A life eclipse.

But who was really dying? The young katydid? Or was I not, in truth, witnessing my own death—I who had just been so intoxicated by tenderness and the joy of the light? It had been my own life that had quickened within me as the light of the young leaves reached into my own being through that little body with the fragile mandibles and the black button eyes. And the question shot painfully through my soul, pierced as it was by the vanishing sun, itself a point of light among other dying points of light on these immutable hills: Who was "I," actually?

I had no answer.

I was speechless and sad. And yet, at the same, resigned in a strangely serene way.

Just as the sun's loving energy, transmitted through the plants' bodies, nourishes us through our life and is so very much like us—warming, loving—something within me welcomed the death as I encountered it that day in the katydid. Death was here, in me, just as life was and had always been—and there it remained. And this dying contained more than just gloominess. The transparent sprightliness of the katydid was a facet of these mountains, as was this moment of being consumed just a few minutes after that happiness. As was the process that it had now begun, of being enlivened again as part of another body and another life-form by serving to nourish it.

I had the intense feeling that I was born of the leaves and the light, just like this animal, and that I would be lost in them again someday. I was no different from any of it—from the cricket or the evening in those mountains. The transparent being that I looked through as though it were a crystal ball was a mask of the whole that I could recognize only in it, in that individuation. I saw that this whole included copious shares of horror and of happiness.

The shock only lasted a moment. Then I walked on, purposeful, distracted by a thousand things, harnessed to an industrious life.

Without my years spent in Italy, I would never actually have understood some of the important aspects of my own aliveness. Without the sunless weeks beneath the smoke of wood fires in the granite air of the cold, gray Apennine mountains, without the unquestioning goodness of the orchid meadows in May, without the odor wafting off the dry rocks, smelling of heat and aromatic herbs and gifted with immeasurable clemency on an early summer afternoon above surf in Bonassola, these feelings would have remained submerged in the depths of my history: delight and fear, so elemental as to be often painful. And this pain is essential, as it enables me to be alive. This is my lesson from the "glory of expanded noon" on the southern sea, from the endless light of the Estonian summer, from the silent starry world of the wintertime Apennines.

Living means learning to die

The universe is not purely gentle. It is just as deadly as it is gentle. And it can only be gentle because it is deadly. It can only be gentle insofar as its gentleness constantly puts up a fight against death. This is the message of erotic ecology, one that it sets against Darwinism, liberalism, and all of the dominant goal-oriented ideologies, all of the ideologies of efficiency, of combat, of war as the father of all things.

But even if it is difficult, the message of erotic ecology does not claim that life actually consists only of cooperation, that existence is actually ecstatic and healthy, that death is ultimately an illusion. Symbiosis instead of competition! No. Symbiosis and our desire for it are real—but oppression and violence are no less so. The birth of life is a genuine power in this universe. Death as well. Light and darkness—both are essential if anything is to be created.

As a result, we will only enliven ourselves if we are able to see both sides. If we learn that both of them can only exist through the other. And that when deciding how the mixture of the two should look, the hedonistic pleasure of the moment cannot rule the day, nor can the wish for the fewest possible discomforts, nor the fear of leaving one's own comfort zone—only the objective matters. And that objective is to increase the possibility for more desire for being, to increase the level of freedom, to love aliveness. This—not the factual bulimia found in schools—must be what composes the learning processes of our youth and young adulthood, but we have forgotten this.

The message of a world in which all things interpenetrate and consume one another in order to affirm and to realize their desire for more being is not one of universal nonviolence. On the contrary: This world is a tragic universe; tragic precisely because it is creation in the genuine sense, because the act of creation is being consummated in every moment, because this world is inwardly alive and is therefore constantly producing real deaths every second, one of which will assuredly be ours someday.

The great project of modernity for the last five hundred years has been to hide this death, or even to abolish it by means available only to human beings. And to be sure: As I sit at a comfortable desk in a European metropolis of the early twenty-first century, this project seems nearly to have succeeded, or at least to be on the right path. But from the viewpoint of the southern part of the world, the manmade deserts of once densely forested Ethiopia; from the viewpoint of the Chinese sweatshops; through the eyes of the last remaining specimens of one of the many hundreds of species dying out every month; or even just in the feeling of drowning in some insipid pastime, it no longer seems as certain that death has been efficiently defeated. It almost feels as though obliteration always finds a little backdoor through which it can slip and fill its bushels. If one builds a dam, hugely massive and mighty, to hold back death, it will only collapse in more dramatically spectacular fashion.

The humanist psychologist Sam Keen says: "Our heroic projects that are aimed at destroying evil have the paradoxical effect of bringing more evil into the world. [. . .] The root of humanly caused evil is not man's animal nature, not territorial aggression, or innate selfishness, but our need to gain self-esteem, deny our mortality, and achieve a heroic self-image."[9] What makes life into a site of dying, a site where birth no longer takes place, is solely our denial of the fact that death is a necessary part of life, and that indeed our own death is essential to our own life.

This death, which is as much a part of life as birth, does not confront us for the first time at the physical end (even though that death will come knocking at the door someday). It includes every little death of parting, of uncertainty, of nakedness, of helplessness and defenselessness. It is every moment in which I am not in control but allow another the chance to speak—every moment that I go without in order to bestow my share upon another. Broadly conceived, death comprises what I require for life, but am not—things that, if I do not find the proper balance with them, will kill me just as swiftly as my holing up inside myself, inside my immoderate needs and my frenzy for control.

Just as cells can only survive by casting off their substance and building themselves anew in every moment out of the flesh of other beings—in other words, by cultivating death in all of their actions—we too will only be enlivened when, in the face of the terrors that this life holds in store, we discard our illusions, break open the emotional armor that supposedly protects us against fear but in truth simply causes our souls to harden into prisons.

"To philosophize is to learn to die," wrote the French humanist Michel de Montaigne: To philosophize, to understand the world, is to understand in the innermost core of one's own existence that death is waiting within life. What is left to us, if we accept the reality of life, if we wish to be enlivened, is nothing other than the "conscious choice to abide in the face of terror."[10] The erotic view of ecological interrelationship seeks to express that fact alone: At the heart of erotic productivity lies the act of abiding death.

Montaigne followed the great Socrates, who in ancient Athens had also taught that we must completely and fully accept our frailty, because it is reality, and it has never done anyone any good to ignore reality. Marshall Rosenberg, a psychologist and the inventor of nonviolent communication, puts it a little differently: "If you deny your needs for fear of scandal, you will always end up paying for it someday—as will everyone else affected by it."[11] Being aware of your needs and not deceiving yourself about them—this means learning to die like the amoeba that does what it has to do in its microscopic environment, like the mother fox who cares for its pups until she perishes, like Muir's water ouzel: Do what must be done to make a place for life, regardless of what the consequences might be to yourself.

To recognize reality for what it is, and to recognize death within it—this is nothing less than the heart of the erotic: "We are the theater of the embrace of opposites and of their dissolution,"[12] writes the poet Octavio Paz in his book about love, where he dives into the character of attachment and aliveness more deeply than almost any other author before or after him. We are "resolved in a single note that is not affirmation or negation but acceptance."[13]

Learning to die means seeing reality without nudging it in some pleasant direction. That alone is what it means to really see. And that alone is what it means to be sculptural, to be creative without having to be ashamed of your imperfection. That alone is what it means to be wild, wild in the sense of an animal who does what is necessary, wild like the whole of the natural world, which manages its own needs with "mindful[ness], manner[...] and [...] style,"[14] as the poet Gary Snyder says.

Acceptance means being able to do two things: be present in the here and now and work for change with all your power. A living organism does precisely this: It attends to the current moment, completely in the now, but it does so only by continuously getting away from the matter that had composed it a moment before. The lesson of this ambivalence is not dull biologism, but the ecological array of living things. It is the ethics of a practice of being wild, of being a living thing. "Acceptance means striking a balance between ideas that might seem to oppose one another: you're fine as you are at this point in time *and* you have things you want to change; you're not to blame for what you experienced as a child *and* you are responsible for creating the life you choose now," conclude the authors and psychologists Kimberlee Roth and Freda B. Friedman.[15]

Emotional suffering as fear of reality

Reality is what is. Admitting reality means accepting what is—the ineluctable aliveness that carries unavoidable death in its center. The psychologist and humanist Ernest Becker sees the deepest error of our civilization in the constant narrowing and destroying of this space out of a fear of death. And this fear of death, as Becker's colleague Abraham Maslow observes, applies not only to actual dying, but also to the chasm it creates within us: to what is really there, threatening obliteration. The fear of the depths of our own I, Maslow says, often accompanies a fear of the outside world.

The desire inherent in creative Eros seeks to break through this fear. Eros is the marker of life that moves through death in order to become itself. This observation has an emotional, psychological import—but also a material, embodied one whose sense is connected solely with following the physiological needs of a continued existence. Eros, according to Becker, is "the urge for more life, for exciting experience, for the development of the self-powers, for developing the uniqueness of the individual creature, the impulsion to stick out of nature and shine [. . .] the urge for individuation."[16]

In his proposed solution, Becker follows a line of thinking from the German psychoanalyst Otto Rank. Like C. G. Jung, Rank belongs to that first generation of Freud's students who deviated at significant points from his rigid thought, armed against all forms of external criticism, and were pitilessly cast off at some point by their teacher. Rank committed the mistake—unforgivable in the eyes of the Viennese master—of thinking that the concept of an innate death drive was nonsense. Rank asserted that humanity's lust for devastation, its evil, did not result from a destructive impulse built into our biology. Instead, in Rank's view, destructive impulses are impossible to keep in check because we constantly act contrary to our drive for aliveness. And we counteract it because we cannot accept that this aliveness demands that we look death in the face, that we directly approach death as a central element of living reality.

For Rank, as well as for the somewhat younger doctor and psychoanalyst Wilhelm Reich, modern humanity's widespread refusal to surrender anxious self-defense in favor of aliveness (both one's own and others') is the true cause of evil. The "armored" person is the problem,[17] the person who is not open to others, who does not actually want life as such (life that includes death), but places his or her own survival above all else, no matter how narrow and ingrown it might be.

Unlike Freud, Rank does not believe that the reason for psychological suffering lies in our repressed sexuality—no, it is in our repressed aliveness, in our fearful deafness to the voice of reality that demands we accept death in every moment and risk it for the sake of life. Not necessarily bodily death, but certainly the death of the brusquely independent and manipulatively controlling ego that conceives of the world not as a locale emerging from the relationship between creative novelty and unimagined identity, but as a war zone in which there can only be one victor and a host of losers.

Self-interest, egocentrism, the avoidance of genuine closeness, manipulation and control, the exploitation of the weak, the act of basking in the glow of one's fragile sense of self-importance, the sometimes torturous, sometimes self-aggrandizing inability to be alone in our clamorous culture in which so many are suffering (without ever admitting or even knowing it)—all of the characteristics that we identify as the symptoms of our own and others' neuroses are caused, in Rank's eyes, by the fact that we seek to protect the I from all forms of transformation and openness to others, no matter the cost. All desperate seeking for self-realization and ego-affirmation follows this protective mechanism: It longs for the triumph of the closed-off and safeguarded sovereign I, for invulnerability. But there never can be such a thing as that ego of the fearful dreamers, because all life is a creative process in which a sensitive identity can only develop by opening itself constantly to new things and thereby continuously giving up a little part of its self.

For Rank, emotional suffering that seeks a scapegoat is thus not a sickness that must be treated in therapy, but a form of misconduct that a mature person can put aside, having gained insight into what must be done. This is precisely the form of maturity that our culture so desperately needs now, during this sixth wave of species extinction, during this orgy of self-aggrandizement. We are lacking the manners and grace of such maturity nowadays. But we can achieve it anew. We simply have to give ourselves a push and be brave once more. Our path into maturity, which is also our path back to enlivenment, leads us toward a regular practice of laying aside the armor that offers us no protection, but only fetters us.

"Free is the one who is ready to die," writes the Danish author Karen Blixen in her great Africa novel. She writes it because she is an artist and knows something of the painful mystery of creation. Creation, whose freedom is indeed no trifling matter. Unlike animals, unlike the tragic transparent green katydid, unlike Muir's water ouzel or Rilke's gazelle, we do not have an inborn mastery of death. And we thoroughly teach our children to forget the little part of death that they can already master—or better yet, the part they already know. Learning death as an emotional attitude would thus be the task of our culture—as it once was in some cultures. The thoughts of Plato's Socrates and the cheerfully stoic views of Montaigne are both examples for us. Nowadays, they could perhaps have a function, leading us. Welcoming the Eros of living means nothing less than being ready to die, accepting the unavoidability of breaking down. Only then will we fully be our bodies, and at the same time, much more than them. Complete creatures of matter and complete preservers of the freedom that constantly entangles matter in creative overflow.

Citing Rank, Ernest Becker writes that only if we submit to the greatness of reality at this highest of levels, where it is least misused, least abused by others for their own ends, only then can we conquer death. Not by evading it, not by armoring ourselves against it, not by rationalizing it away with research, technology, and economic growth, and not by pinning it on others. Rather, by "combin[ing] the most intensive Eros of self-expression with the most complete Agape of self-surrender"[18]—the surrender of one's own goals to the cause of aliveness as such.

The complete obsession with deep inner goals and the complete surrender to the purposes of life as a whole—both of these, combined in paradoxical unity, can leave fearful death behind and turn it into a tool of productivity. This is the way of the water ouzel, its wise breast bedecked with icy drops, who thoughtlessly does what must be done and thereby becomes, in Muir's eyes, a crystal ball in which the facets of existence are wordlessly interconnected, the blooming flower of the wild brook.

Yet in our epoch, there is hardly an imaginative infrastructure in place for such thought. Becoming aware of life's creative reality plays a minor role. For the overwhelming majority—particularly for those in power—the natural world is an economic resource or a factor of economic stability to be grasped by purely technological means. At best, it might play the role of a pleasant diversion in one's private life. Such an experience is, of course, detached from the understanding of the deeper reality of which my actions are a part. But life will only be able to gain a stronger foothold in our politics and society through the encounter with death. Erotic ecology thereby conceives of itself as a first attempt at the infrastructure of a world that experiences itself as deeply alive.

The answer lies in your eyes

The great exercise of learning to die, perhaps the most urgent exercise that our culture needs to practice so that it might again reflect on what truly is, cannot be advocated under the protection of "ultimately correct knowledge" or from within the fortress of a saving declaration, "Oh, that's it!" We do not come away clean. Anguish is a part of it. When you seek to do away with it, you only beget more. The Nietzschean Übermensch has failed so dramatically and in so many ways over the last 150 years that the glorification and heroization of death is no longer appropriate. It is among the powerful's instruments of seduction, their siren songs, falsely promising an eternal life that, ultimately, will belong to them alone: return on investment and jihad.

No. There is a world of difference between the glorification of death as another variation of the delusion that one can come away cleanly, and the sober recognition that this world is composed of light and shadows and that everyone who seeks the light must also welcome those shadows. Learning to die does not, in fact, mean learning to kill. Nor does it mean learning to be killed. Rather, it means: serving the unfolding of enlivenment, of the self-in-connection—one's own self and thereby the self of the whole.

This is why it is so central to welcome the right death. The conspiracy to do away with darkness in which everyone in our civilization has taken part for the last two hundred years, the human salvation of the world from evil by means of technology and laws, always had the advantage of wishing everyone the best: that is, wishing them life, eternal life. But we have now seen the extent to which this wish produces its opposite. Wishing for life at any price continuously calls forth death—the death of other people, other beings, the extinguishing of languages, ideas, and, worst of all, possibilities and degrees of freedom.

The secure bank is an illusion. But preaching the readiness to die with Greco-Roman virility quickly brings about a new dilemma. Doesn't it sound eerily like blood, death, and honor—"dulce et decorum est pro patria mori"? Because the bloody trail of sacrifices to ever-new variations of the glorification of death runs throughout our cultural history, we find it difficult to overcome the reflex to shy away from the simple fact that death is real and truly belongs to the principles of being. It is politically incorrect to speak about necessary death. Too many people have presumed to make decisions about whose death might be necessary.

But others' hubris should no longer compel us to shut our eyes to reality. Nor should the fear that we are deceiving ourselves. For whenever we speak of death, we are not speaking of another's death, but of our own. This is the myopia of our culture: its failure to see that being alive means preventing the death of others, as much as possible, and thereby ceasing to fear our own. This is not at all immoderate, but is actually a form of moderation (I speak in more detail about this "middle way" in chapter 7). It means accepting the deep understanding that it is not about me.

In the understanding of aliveness that views life as an erotic phenomenon, as a phenomenon of being touched and being in relationship, it is not about welcoming death and then declaring it a remedy. It is not a question of nihilism. Nor indifference. Nor technocratic stoicism. Nor heroic romanticism. It is not a question of realism, which so often conceals an unimaginative cynicism. On the contrary, it is a question of equilibrium, of the living midpoint between glimmering light and devouring darkness. It is about despairing over the fact that suffering is unnecessary and then, for that very reason, choosing not to avoid our own suffering, for life's sake.

To abide death does not therefore imply a nihilistic declaration, "What's the point!" The attitude that truly follows aliveness's call to the end knows to distinguish between death as a slave in one's own existence and death as a wild animal that freely follows its own needs. It is still appallingly difficult for us as human beings to make the correct determination here. Recognizing the brokenness and contradictoriness of life does not mean (more contradictions!) acquiescing to it and accepting it and, in the best case, transferring it into the "spiritual worlds" of meditation practitioners and philosophical theories. Recognizing it means constantly taking action against the rupture and thereby learning that it will not be closed. This world is broken beyond help and yet, because of the littlest of gifts, it becomes infinitely and immediately better in perceptible ways. This world is tragic, because tragedy is the energy from which the living creates its plenitude—but actually, it is only living because it runs up against this tragedy in every moment and places limits around that tragic element through a further gesture of transgression, through a new, productive imagination.

Accepting your own death therefore does not mean persevering in abusive relationships because that is the only way, because all relationships end up there anyway—it does not mean throwing some of your own life into the void. Being healthy in the form of learning to die means no longer fearing your own death so that you can stop sending others to their deaths—and so that you no longer perish in some false death, a demise of imprisonment or depression in some airless hideaway that you dared not leave because then the light of day would shine painfully on your nakedness.

The false death follows the act of perishing away in a prison of your own fear (which only buys its life through the cruelty shown to those who attempt to make it through life with less armor). The "proper death" is the one that is not sought. It has nothing at all to do with insipid heroism. But if you really want to live, you must take death in the bargain, because it is inseparable from the struggle, for the sake of aliveness, to unfold your own creative possibilities in a world that is full of other creative beings.

This, then, would be a first approximation of the existential ecology of death: In order to hold on, I must let go of one moment so that I can exist in the next; I understand that every stone whose uniqueness moves me to tears does not belong to me, although it is made of the same stuff, although it, like me, is unique and completely itself. An abyss lies between us, and this abyss is death, my complete separation from the world—and at the same time, this abyss enables the uniqueness of all things, and also my own specialness, the very specific flavor of my idiosyncrasy in this world.

In order to let in the living world, I must be completely vulnerable and must learn to be truly defenseless, in a state of utter precariousness, like all of my cells are from moment to moment. I must exist in absolute uncertainty in order to completely perceive reality. This is nakedness *in extremis*, the nakedness of the animal, the nakedness of the world itself.

The psychologist David Schnarch writes: "Our problem is [. . .] our unwillingness to meet life on its own terms."[19] Meet life on its own terms—that sounds like empirical science, like genes and instincts. But Schnarch does not mean it this way. The terms that he is speaking of are the principles according to which living subjects unfold and become themselves by depending completely on others; these terms are the foundation upon which selves are established, maxims according to which vulnerable bodies occupy their own positions in the world—and our position is in fact our emotional perspective, our experience, what we go through. These terms are the principles of the conditio vitae, the condition of being alive that we share not just with other human beings. Gary Snyder says: "To be truly free one must take on the basic conditions as they are—painful, impermanent, open, imperfect—and then be grateful for impermanence and the freedom it grants us."[20]

These terms are not at all compatible with the Darwinian laws of mercilessness; they are not part of the "eat or be eaten" worldview that we have all subliminally adopted for reality. Accepting life on its own terms does not mean following the rules of combat or submitting to a game in which there can only be one winner. The terms of life are different. They call upon us to not abuse another person, other beings, or the rest of the inanimate world for the purposes of exercising control over our own fate. Rather, they indicate that we are both fully free and completely dependent on our fragile bodily and emotional constitution, and that in order to take on the responsibility for this freedom, we must risk that fragility without blaming others for our breakdown or recruiting them for our protection.

The terms of life indicate that we should try to be as alive as possible and in so doing recognize deeply that we are altogether mortal. Indeed, that in order to become more alive, we must die over and over. We should never try to disregard this provocative fact. Accepting the terms of life means recognizing that only we can support one another in this misery. That this is the only way that love will find its way to us—when we do not look to it for protection from loneliness and death.

So that would be it then. Accepting tragedy. Not running away from it, not deactivating anything. This is the error of modernity: It believes it is possible to prevent the tragedy. Yet the attempt to foreclose it only makes it worse. Masking pain increases it. Deactivating feelings invisibly intensifies them elsewhere. Ignoring one's own suffering causes others to suffer. Attacking the brokenness of the world at the root leads to violence and oppression. Life at any cost can, in extremity, lead to murder. And at the same time, we must do everything we can to save life from unavoidable destruction, no matter the price.

How can these paradoxes ever be resolved?

The answer lies in your eyes. The answer lies in the radiance with which you greet my presence, because it gifts to you a share of aliveness that, echoed in your gaze, welcomes me. The answer lies in the fact that I will do everything to prevent your death, and that this "everything" includes the possibility of my own death. I cannot ward off the biocentric tragedy. But I can live it to the fullest, can make myself into its embodiment. I can take on the responsibility for it. I can do what is necessary for you to live and for me to live. I can take on good will for your life, as I have for my own. I alone carry the responsibility for myself. For my courage. For my death.

PART TWO

YOU

I effuse my flesh in eddies, and drift it in lacy jags.
I bequeath myself to the dirt to grow from the grass I love,
If you want me again look for me under your boot-soles.

WALT WHITMAN[1]

— *chapter four* —

TRANSFORMATION

Every separation is a link.
SIMONE WEIL[1]

I remember an evening from 2013, when the summer had just begun. It was a still evening, an evening without wind, when the day's warmth seemed not to want to leave the air, as though the mildness of the coming summer were flowing into the world and filling it to the brim in the darkness. It was also a special evening because in its solemn stillness I heard the nightingale sing again for the first time. The nightingale, that magical being of transformation whose voice can enchant the whole world, as though everything were suddenly transformed into some new material, as though things were made of chiming glass and the air of red velvet cloth beneath an immense bell.

The nightingale, oh! that wonderful bird, which I could write a whole new book about every spring. The nightingale, that tiny creature weighing just a few grams, practically immaterial, purely voice. On that evening, its world-altering power overwhelmed me with a feeling of wonder and gratitude. I also felt melancholy for the already certain transience of our encounter. My heart pounded as I grasped how much I loved this little bird, how much my soul was attached to it, how much my feeling was changed by the touch of its tones.

And my heart beat faster as I understood that all emotional encounters inevitably transform us. All relationships are transformations that leave both me and world changed by one another, encounters in which one penetrates the other and leaves it altogether different than it was before. Everything changes when we engage with it in emotional contact. No encounter leaves us the same. We cannot be neutral. We are always already swept up. What we see or hear changes our perception—and our new way of engaging with things causes a change in the way we make contact with the world. We are never the same from one second to the next. We are constantly becoming—and the place in which we live changes along with us.

In that evening's endless minutes, in the sonorous, atemporal exile that this little bird afforded me, it became clear to me that this, too, is a power of physical touch. Transformation is thus another part of erotic ecology, determining all of our relationships in life. The living world is a world comprising relationships, not objects. When we are feeling and perceiving, we are not like stones lying on the ground, but like droplets of water caught in a spider's web, slowly sliding along the shimmering silky threads as the morning air grows warmer, transforming eventually into invisible vapor.

We too—and with us, all organisms on this Earth, from the tiniest cells to the blue whales and the rain forests—constantly drift here and there as we form attachments to other life-forms: by speaking with them, establishing friendships with them, understanding something through them, satiating ourselves on them. In these relationships, we become what we could be but were not before. In these relationships, we transform alongside what we perceive. And this capacity for transformation is what makes it possible for us to experience anything. Comprehending something means transforming. Learning means becoming something or someone else.

That same evening, I walked with my dog along the dark street. Instead of going back into the house, I stopped and listened. For a long time, I stood before the nightingale and listened to how it expanded the night into a great dome and coated it completely with its voice. It was quiet enough to hear it fully, there in a thicket on the far side of the tracks. The dog scampered about as I listened, sniffing here and there. She moved gently through space that was both dark and illuminated with birdsong. It was so quiet that it seemed as though the Earth had paused in its flight through the blackness of space in order to offer this nightingale a perfect stillness, that its song might unfold into the deepest darkness and thereby fill it, somehow, with a new form of light.

For me, that summer spent mostly on a quiet side street of the Westend neighborhood of Berlin became a record of transformation. I perceived the world as it transformed, as it gained volume and lost mass by the second in a frenzy of growth and unfolding. I greeted first the blackbird and later the nightingale, returning after a longer absence. I counted the scanty snowdrops in the garden. And later in the year, in amazement, I inhaled the fragrance of the lindens in which the bees buzzed domestically, that childhood smell of lindens that is so sweet and so hopeful that it can make you dizzy with happiness. I have seldom experienced the overlapping phases of spring and summer more intensely than I did that year. And yet I had moved away from an idyllic house at the edge of a conservation area and into the city. But maybe this is precisely why. Transformation hung in the air, it struck me more than ever before, I eagerly savored its power to stir the world, and I experienced every second differently.

But perhaps there was something else: I transformed the world itself by perceiving it. There were two somewhat neglected parks in the area. I experienced their open spaces as a refuge for my soul where I could participate, to my heart's content, in the general transmutation of light into desire and fullness. I strolled through a landscape that had been endowed with the same capacities as my soul, one whose dramatic manner (everything was still shooting upward, but soon it would yellow, then quickly disappear) spoke to my inner emotional roller coaster.

I don't believe that I saw my tangible soul there before me, nor do I believe that "nature" is always soul. But there was something that connected me with the living landscape surrounding me, a common principle of existence that I could recognize and that enchanted me as I perceived it. I transformed myself and everything around me, and yet the thistle flowers were still themselves, as were the blades of sweet grass, the field mice and the wood ants, the woodpecker and the song thrush. They remained themselves and were therefore ultimately incomprehensible, but they opened themselves to me at the same time. It was an encounter, and in this encounter the world invited me to be at home in it. I transformed these beings through my gaze, perhaps also through my joy, and that also made me into someone else.

I reveled in changes great and small. Gently, I was lifted high on the long waves of the perennial changes in vegetation, but like the tingling caused by the unanticipated touch of a stranger's skin, I also enjoyed the lightning-fast, unforeseen renewal brought on by the ever-changing weather. I exalted in sudden sunshine, in lightning storms, in velvety gray. I was hooked on rushing out the moment the rain had stopped, when the sun began to dry out the world anew. Again and again, I got drunk on connecting with that light, slightly incredulous fizz with which the landscape would reform itself after every shower—steaming and glistening, as though each quantum of light were broadcasting a message that it was ready for every imaginable form of creation.

I ran out into the alley of linden trees where I lived, tugged into the light by my little black poodle mix, that muscle of joy who greets the world with mystical ecstasy every time, who seems to never be tired and hasn't a care. So often during that early summer, I would simply jump up, leaving my work, an unfinished half sentence broken off in the middle, because this was more important. One early evening, for example, a heavy cloudburst from the west had darkened the sky to violet and then, after moving off, left the trees and houses and glittering streets behind in a composition of gleaming crystals. I hurried downstairs—barely any drops were falling now—and watched as the sun, slanted and vespertine, gingerly cut out the foliage of the trees from the mist, making them visible again. I listened as the blackbird began to sing in the still sparse, barely green larch, triumphant and impassive, and another answered its call from a tall pine, pallid green with an almost orange trunk.

In those weeks, I experienced the opposite of the fact-based fascination of the particle physicist, whose measurements show that the world is just empty space between the tiny probability clouds of the atoms. No, it was nothing like that! Here, beneath the soundless thunder of the evening in my unassuming linden alley with its two forgotten and wildly overgrown parks, you could see just how full the world was—practically overflowing at the brim. Here, you could see that the world was an unimaginably dense web, a network of calls and responses, of hastily inhaled breaths and indiscriminately exhaled fragrances, of oscillations and consolidations, of sobs and laughs as tears are dried, of dense humidity and ephemeral vapor. In this intertwined root network, it was a home overflowing and saturated, even as the indisputable analysis of quantum mechanical equations showed that it was almost completely a vacuum. In truth, the biosphere is actually both simultaneously—billowing lungs and speechless stillness, fathomless between two breaths, between two hands that fleetingly engulf each other in the act of being rushed past one another.

I remember that evening in May because the transformation took place so completely, because the clouds had been so darkly black and the rain so extremely heavy—and then everything simply lay there in unexpected, unprecedented splendor. I walked through the park under the dripping trees. A homeless man in a tatty sleeping bag had retreated under the steel S-Bahn bridge that cut through the green space on the way to the Olympiastadion. He lay in the sand and did not look at me as I walked past with slow steps.

The water dripped from the trees in a stillness in which everything seemed to dissolve as though in a cleansing tincture—the nightingale's melody, the blackbird's fluting cadenza—into which smells seemed to come loose, making it hard to breathe because of the fragrances, as though the leaves themselves might evaporate. I listened to the sound of steady dripping under a sky washed immaculately clean, a dropping sound—"as though the leaves of the garden trees *copied* a steady shower," as Samuel Taylor Coleridge once noticed in a similar situation. In the break following the rain, everything was still wet and yet already dry. This, too, was a moment that was already a different one, an endless instant, a transformation slowed to the point of imperceptibility.

A few days later, I went on a playful exercise with my daughter Emma in the Grunewald. We didn't run; we hopped with long leaps through the balmy air and periodically had to stop, convulsing with laughter. For a while, I carried her on my back, then she chased me again, and then we both watched in fascination as our poodle threw one of last year's chestnuts high into the air and prodded it with his snout, as though it were a ball and he a trained seal.

I tried to use the new leaves, still almost completely transparent, to explain to Emma why I thought that you could talk about "the gold of the new green" when describing fresh foliage. At first she didn't listen, and then she laughed at me. But then I understood that she was still gold herself and that she didn't need to understand all of that—she needed just to be it.

Transformation as translation

All perception is a form of being touched, and all touch is metamorphosis. Three chapters ago, I described how the oceans perceive the moon by stretching toward it in the tides. The Earth transforms itself through the moon, just as the moon experiences itself in a very particular way for as long as it is connected with the Earth, because the Earth's mass is what holds that heavenly body in its orbit around our planet.

When we consider the world from the perspective of metamorphosis, we arrive at an altogether different picture than we do when we analyze it in terms of its causal relationships, which science up to this point has primarily done. Suddenly we discover how much connectedness there is—and we perceive the separation less. For example, why shouldn't we be able to say that the Earth transforms not only the moon, but also the sun? On its surface, the Earth converts the sun's energy into a skin of tumescent flesh. Sunlight transforms through photosynthesis into the bodies of the plants. With the whole surface of their bodies, the plants troll the sky for energy and thereby become the growing matter that is at the beginning of all food chains and anchors the "great chain of being" in light. The biosphere is sunlight become flesh. One could say: The leaf's green translates the sun's energetic gift into body.

So that the plants have access to this gifted energy, every impulse to grow within them is directed at the sun. Through the movement of tiny granules on pillows made of delicate membranes, gravity cells help the vertical axis of the plant align with the center of the Earth; light-sensitive pigments use hormones to instruct the growing tissues to direct the leaves always toward the light, which is what causes plants on the windowsill to grow so rampantly one-sided: It always seems as though they want to get out. In a certain sense, the plant "sees" the light—even though it is a sensation experienced on the outer surface and the perception is not an abstract cognitive act, but expressed in the altered form of the growing plant, reaching for the sun.

Animals developed through the window of light that was opened by the plants. Anything that lives on plants must be able to see the things that provide it with sustenance. Once oxygen production was instigated by the first blue algae, all other biological possibilities had to occur within its world of light. To this day, we are heir to this Siamese interdependence: We say that a happy person beams. And of course, people still retreat to the green of the plants in order to feel at home. Just as the plant's color empirically soothes our eyes (the eyes of a former plant-eater), so, too, does the quality of transformation that it imparts vibrate as we look upon it. In the plant, we see something that spontaneously promises life. In the luminance of its blooms we encounter a reflection of the sunlight, an appearance of potentiality made thing, which is the paradox of creation itself: the paradox that, in order finally to become itself, nothing may remain as it is.

The plant can only grow when it systematically lets go of itself; it can only become matter by getting involved with something (light) that is entirely other than matter, an excess of which has the possibility of destroying it. The plant is not the stuff that composes it. Yet for a precarious moment, it can bind this transitory stuff into something, the greening body, that proves to be lasting and extremely tenacious. But this identity, this "self" of the plant, is not actually the matter that composes it but its translation into a success story that could be over at any moment.

One could say the plant "understands" an aspect of the matter that it takes in by translating it into its own body. The plant translates the world into body and thereby reveals further dimensions of the world that had been previously invisible. That the world can bloom, for example. Or could one even say that the plants' blossoms are the echo of the sun and its streaming light, in the form of a vulnerable body?

In this world of mutual reflection, something else comes through: We, too, understand more about the world's potential in the act of transformation. When we see the plants blossoming from the Earth every spring with new energy and childlike confidence, they make something visible for us that was hitherto unrecognizable. And since plants are life-forms like us, they are actually making comprehensible a dimension of our own existence that we do not often see clearly. In a sense, they demonstrate the terms of life in their own bodies. They show us, as the poet and philosopher Friedrich Schiller believed, that necessity can be bound up with freedom, and that beauty can result.

Our body translates our constant material contact with the rest of the world into meaning. Matter that seeks to maintain itself in some particular, improbable, animate form—as do amoebas and hummingbirds, blue whales and tardigrades—interprets the world of stuff and of energy transfers into a world of meanings and subjective experiences. Matter that seeks to preserve its own existence, as metabolizing bodies do, thereby transforms the exterior world into an interior world, constantly creating meaning out of touch. This is a fundamental aspect of erotic ecology.

Meaning is the result of the translation of a material outer world into a meaningful inner world. We experience this inner world as feeling. Feeling is transformation, the transformation of another's presence into one's own experience. Feeling means becoming another person because of another person. Feeling means that one piece of the world "folds in" another, calling forth an order that contains both and neither, because it is something altogether different.

I am constantly fascinated by how much this biological viewpoint reflects the perceptions of modern physics. There, it is recognized that every measurement alters what it measures, even though this effect can only be observed when the thing being measured is very, very small, like the size of a single atom or even just an electron. But in that case, the momentum of the measured object cannot be determined unless the researchers also establish their own position. The little particles that they must then emit to measure the other little particles interact with them and change them, or even destroy them. It is similar to what happens when a hunter shoots small shot at a bird in order to figure out its location. Afterward, the bird is not the same as it was before.

Our feelings are the same sort of transformative reflection of the other. Our inner worlds are so irrevocably changed by these reflections that nothing is as it was before. Because of this altered inner world, our actions also change, which in turn affects the outer world. For a long time, such a connection between outside (the body's biochemistry) and inside (our subjective experience) was considered absurd in biology. But nowadays, molecular biologists are discovering how feelings, the pure "inner dimension," alter even the genetic material. Emotional experiences can cause the conditions of our DNA to change, and this alteration will then be passed on to our descendants, just as traumatized people bring children into the world who themselves show traces of similar shocking experiences and who have a greater tendency toward fear and uncertainty written into their genes.[2] The outside is translated into the inside and then reinscribed—a magical echo chamber, a cabinet of mirrors in which nothing escapes and new pictures are constantly drawn.

Incessant creation: existence as imagination

As I made pilgrimages through my little West Berlin parks, I witnessed this constant act of translation. I witnessed inner worlds unfolding—but I could only witness this by perceiving these inner worlds from the outside, as the bodies of emergent flowers and grasses. I could only perceive their inner side—their desire for more being and their fulfillment in growth—by feeling this desire and this joy myself; indeed, perhaps only through the act of sensitively formulating it, writing it down, transforming it into words, and sharing it.

I can only come into my own in community with other living matter. My developing self depends on the gaze of others, on the warm gaze of the plant that is entirely eye, feeling the sun's light all across its surface. Nature is such a part of me—namely the part that sees me because it is not me, and that therefore understands the part of me that I am, but lack: my psyche, which is plant. I remember how, as a young man, early in the summer in Tuscany, I walked through the night between the fireflies and the stars and thought that I would have to be in love in order to have a feeling adequate for all of this. It seemed to me as though I myself would have to be more beautiful in order not to become lost in all that beauty, in order to respond to that desire for life with a desire that was able to capture it. I walked through what the poet Rainer Maria Rilke had called the "world inscape."

Metaphor in the "syllogism in grass"

In what follows, I would like to try to establish the connection between the transformative power that is part of our living experience and the imaginative power of poetry. It seems to me that the poetic mode of perception and our experience as life-forms are more intimately connected with one another than our technological culture has long claimed. Perhaps it is even true that at the beginning of our lives, as babes and little children, when everything seems so magical and shining and new, we only have poetic experiences: quavering, inspiring, illuminating, connecting us intimately with the world.

According to this point of view, poetry—an expression that conceives the world verbally or artistically, but not through explanation—is the appropriate instrument for experiencing the erotic. Indeed, the erotic (in the broad meaning that I have offered in this book, as an embodied experience of being on the Earth) might be thought of as the bodily component of poetic experience. Octavio Paz expresses this idea when he writes: "The relationship between eroticism and poetry is such that it can be said, without affectation, that the former is a poetry of the body and the latter an eroticism of language."[3] Both modes of experience are based in the sensory logic of a flesh-and-blood existence, an existence as relationship. Both are intensive experiences of how it feels, as a body and thus as a part of the world, to be in a state of constant exchange with the rest of the world. Both are sensory experiences that cannot be had without a sensing body, because an abstract description is not sufficient—it requires an expression that can be sensed by the eye, the ear, or the skin. A kiss is softness sensed on the skin; a poem is that same tenderness, provided with a body by the velvet of a melody of words.

In its imaginative potential, in the imaginative excess that it appends onto reality, the creative word holds the key to the understanding of what there is. But not because there is no true reality apart from our discourse on it. Rather, because living reality is already constantly reaching beyond itself in continuous, creative transformation. Our words and intonations and metaphors that make us shiver or shrug are part of this transformation. They are part of the body of the world. The world speaks itself.

At this point, I would like to overturn the viewpoint, widespread in the humanities, that we hopelessly overlay worldviews onto an unknowable universe with our "interpretations." Not so. We are able to understand the nature of creative reality precisely because we, like all life-forms, are part of this reality. We are the world. We have both embodied and genuinely creative experiences, and we express these experiences with imagination, creativity, and freedom. We express our being in the world by the same means the world forms itself. Reality is fluid and constantly bringing forth new beginnings—and so is our intercourse with it. A microcosm.

Because we are alive, we can comprehend the living universe. We comprehend it by entering into relationship with it—which is to say, by transforming it and allowing it to transform us. And for us, language serves as an instrument of that transformation. It is suitable for the living world because of its creativity, because of its freedom to voice everything within its expressive spectrum, and because of the fact that it is not a physical instrument of measurement that follows Newton's laws of cause and effect. It offers that "imaginative surplus" that Francisco J. Varela found even in the cognition of bacteria.[4] Metaphorical experience is already the logic of the organic world. Symbolic meaning. Poetic sleight of hand. The boundless paradox. Within that paradox, we can use language to experience ourselves as part of the universe of creative references, comprehending that universe by taking part in it. And language can be anything: It can be every way in which we generate a gesture that has a living quality.

Poetry is the logic of the organic world. It is the only means by which "life recognizes other life," as the German philosopher Helmuth Plessner remarked in the 1920s.[5] It is a channel of understanding by which we receive the clearest evidence of what the doctor, philosopher, humanist, and musician Albert Schweitzer recognized as the deepest core of sympathetic relationships with other beings: that we are "life that wants to live, in the midst of life that wants to live."[6]

No one has gotten to the heart of creative reality's poetic logic more clearly than Gregory Bateson. He compares the classical Greek logic in the "Syllogism in Barbara" (so named after a mnemonic device by students of Scholasticism in the Middle Ages) with the logic of the living, which Bateson calls the "Syllogism in Grass." Whereas the classical syllogism ultimately proves only relationships that were previously known (which is perhaps why some find it so dull), the logic of the living enables a new experience and a previously impossible insight.

> *Syllogism in Barbara*
> *Men die.*
> *Socrates is a man.*
> *Socrates will die.*
>
> *Syllogism in Grass*
> *Grass dies;*
> *Men die;*
> *Men are grass.*[7]

The syllogism in grass is the logic of poetic experience. In fact, the name that Bateson selected distantly recalls a poetic work: the epochal cycle *Leaves of Grass* composed by Walt Whitman in the middle of the nineteenth century. Poetic experience is always bound up with the body. Even when we read the lines of Bateson's argument in the syllogism in grass (which could itself be considered a little poem), we are not just rationally tracing a logical chain—something more happens. We feel ourselves in our bodies, and we feel the touch of the leaves of grass on our skin. We see images before us, recall moments when we encountered grass: as a child, the leaves enormous in our parents' yard, the undulating hair of a melancholy summer meadow, the rustling spread of grass growing in the hard sands of a Baltic sea dune. We live through "grass" and can thus verify the content of the argument. We feel its meaning in the body and imaginatively add in all of the half-hidden memories and experiences that accompany it. We are reminded of already knowing that we are, in a certain sense, grass: mortal, like grass; well versed in swaying melancholy in the wind, like grass; robust and resilient, like grass.

This "logical" recognition is one that we do not arrive at through our "mind" alone, but rather with our whole being. It is possible because we all share something, something that is not simply an incidental characteristic (a color, a scent) but something that defines us existentially, even if it has different effects. We are alive, as the syllogism says, but we must die, and this "we" includes not only other people, but all beings of the living world. Our own experience is supplemented by all of the foreign ways of living and dying that are not our own, embedded in the names of the plants, as though the grass, the tree, the nightingale suddenly became a magnifying glass through which we were finally able to see ourselves clearly.

All of the characteristics that other beings possess are capable of becoming potentialities of our own aliveness and possibilities for our own feelings. The fluttering grass of a prairie swaying in the wind becomes a way of expressing our own desire without having anything to do with it causally. The deft determination of a fox on the hunt lets us experience tenacity and cleverness without requiring that the fox itself be necessarily clever. It is simply a fox, a form of organic potentiality, but through its presence, through its living gesture, we can see the world and ourselves as though we had just grown a new sensory organ.

The syllogism in grass enables objectivity, but this objectivity is not rational—it is indebted to our common participation in the living network. At the same time, it is subjective—subjective objectivity in a world in which subjects are the rule. Since we are all part of a creative, enlivened world, this subjective objectivity is perhaps entirely appropriate. It does not underhandedly deny life by seeking to explain it.

Again and again, Pablo Neruda demonstrates how much poets rely on the body—that is to say, on the embodied unconscious—as an organ of world-fomenting knowledge, thereby becoming (and causing their readers to become) grass, plants, the nonhuman living world. For example, in these lines:

Oh Earth, Wait for Me

Return me, oh sun,
to my country destiny,

rain of the ancient woods,
Bring me back its aroma, and the swords
that fall from the sky,
the solitary peace of pasture and rock,
the damp at the river-margins,
the smell of the larch tree,
the wind alive like a heart
beating in the crowded restlessness
of the towering araucaria.

Earth, give me back your pristine gifts,
towers of silence which rose from
the solemnity of their roots.
I want to go back to being what I haven't been,
to learn to return from such depths
that among all natural things
I may live or not live. I don't mind
being one stone more, the dark stone,
the pure stone which the river bears away.[8]

The symbol is not what it is

The logic of poetry has yet another distinctive feature: It is not true. It is false. People are not grass. They are people, a type of ape, and thereby animals, not plants. This is the reason that some rationally minded people turn immediately away from the syllogism in grass with a shudder. It is not just imprecise, it is quite clearly incorrect. Or to be more exact: The poetic argument is precise *and* false. It hits the mark, but only in the sense of poetic precision, in which another being understands exactly what is meant, even though that meaning cannot be fully described in one clear sentence. Indeed, it is only precise *because* it is false. The poetic experience of the world is precise and false for the same reason that every life-form is both itself and—at its deepest core where matter simply flows through it—not itself, but its own death.

Poetic precision, which always emerges from contradiction, is the basic magic of life. It unleashes the imaginative power by which something simultaneously is and is also its opposite. It is the power that makes my poodle cower when I rise up above her in a towering gesture, even if I am not doing anything. She understands such a bearing in her own body, allowing her to feel—partly due to the help of special nerve cells, the so-called "mirror neurons"—what sort of motor intentions lie behind such a gesture. Even if I am peaceful, I become a symbol of might—just as I become a symbol of the playful companion when I throw myself onto the carpet next to my poodle and bend my upper body low (which immediately results in an enthusiastic dance of the dog all around me). The previously mentioned behavioral biologist and inventor of biosemiotics, Jakob von Uexküll, observed that people in regions of Africa where elephants are native stiffened into a "tree" whenever they encountered one, so that the dangerous, thick-hided creature would trot past without even recognizing them. Here, too, natural symbolism based in contradiction is at work: The people became trees in the perception of the passing elephant, but they remained, of course, people.

Massive weights are moved by the magic of metaphor. It, and not the causal-based technology of the scientific age, is the true earth-moving force. Technology functions only as an effective organ. Meaning gifts something to life—or damns it to death. A warm smile enlivens what it touches. Indeed, for our children, this smile, given in sufficient dosages, is a crucial form of life nourishment. If it is lacking, the soul of the growing person is irrevocably destroyed (more on that in chapter 6). A cold look pierces the heart like a knife but leaves behind no visible traces. All manipulation—the black magic of the soul—is at work in the syllogism in grass: It insinuates but does not speak aloud. But even the white magic of life, the gesture of aliveness, the healing power of accepting relaxation, and also the disposition to poetry, is based on the inner contradiction of the symbolic image.

The poetic expression with which life (as well as death) is both understood and carried forth is based in the same principle of contradiction and incompatibility as the life process, as metabolism itself. And just as this constantly brings about new sensory experiences, so, too, do the targeted linguistic mistakes of poetry contain an imaginative surplus that allows something to be discovered that one had previously intuited but not yet fully known. Poetry is like an arrow shot into the darkness that does not kill, but enlivens what it strikes, bringing it into the light.

In order to illustrate this effect, Octavio Paz cites a remark by the Golden Age Spanish poet, Luis de Gongora, who describes the blood of a deadly wound as "blood-red snowfall." "If Gongora," Paz writes, "says 'blood-red snowfall,' he invents or discovers a reality that, though containing both, is neither blood nor snow."[9] And when the ancient Greek poet Homer calls the water of the Mediterranean the "wine-dark sea," he is speaking a falsehood (there is perhaps no ocean less "wine-dark" than that one), but he gets to the heart of a certain quality of experience—an inwardness, a sweetness, as well as an arcane bitterness—that is more fitting than any "objective" description.

The reality that emerges through such images in the syllogism in grass does not exist—and yet for precisely this reason, it is capable of expressing unique experience. Something is because it is not—in this strange intermediate realm lies the logic of living: the same logic that, as we have seen, ensures that we are bodies through which stuff constantly moves; that we are both binary code (in our DNA) and pulsating matter; that we act with an interest in internal concerns, but within an external world. If we follow this insight, then biology, so long as we take seriously the presence of an ecological erotic in our experience, is no less bizarre than physics, where quantum theory has long since abolished the objective reality of space and time, where the observer and the observed are no longer divided but are irrevocably connected. They are connected *because* they are divided. They are connected by the logic of a paradox.

Citing the thought of his Danish colleague Niels Bohr, who essentially cofounded quantum theory at the beginning of the last century, the physicist Ernst-Peter Fischer remarks that the opposite of a simple truth is a falsehood, but the opposite of a deep truth is itself a deep truth. People are grass.

My friend, Rainer Hagencord, a priest, biologist, and founder of the Institute of Theological Zoology at the University of Münster, took this one step further in a recent conversation: "What you're saying there," he mused, "is indeed a paradox. So it must be true." We laughed long and hard.

Poetry is our wildness

The poetic metaphor is the extreme case of creative language. But this extreme case clearly illustrates how language functions as a whole. Without images, communication would hardly be possible. How could we express joy without a "lightening of the heart," without a "sigh of relief," or "being on top"? Philosophers and linguistic researchers have determined again and again that our turns of phrases and expression are less logical-objective than imagistic-symbolic. Thus, language works according to the foundational paradox of the living world: It establishes connections by conjuring them up, rather than truly embodying them. It is something by not being something. A word is not the thing, but without the word (or some other sign), we can hardly do anything with the thing. Language hits the mark by missing. And this is precisely what makes it into a fundamental instrument of world transformation.

The rules of the biological realm—and thereby the principles of creative freedom whose own existence means something to it—extend not only to the bodies of the animals and plants, but to every living exchange. So we could say: The world in which we live is fundamentally *wild*. We exist through wildness. And because of that, the most civilized world is actually still a wilderness in many respects, although it might be one of the heart or of the mind, and not so much one composed of trees, flowers, and birds. This fundamental wildness of reality includes every self-producing creative connection in which we are enmeshed and which we cannot control, which enables our self-experience while also being capable of any type of expansion through our creativity. If the poetic act is a manifestation of wildness, language is our everyday wilderness. It is a system of creation and relation, which nourishes us, which limits us, which constantly changes. Seen from this perspective, language is not fundamentally opposed to nature, but one of nature's apparent forms.

Poets usually agree here. They are hunters and gatherers in the wilderness of language. They know—just as little children do, unconsciously, as they look for words and build sentences, when they're lucky—that there is a world before it is grasped in words and that the right word does not describe an objective connection, but forges a unique and peculiar one, thereby illuminating the oldest existing things. Language is wild insofar as we can make it our own; it is objective insofar as we can adapt it to our subjective needs.

The poet and Nobel Prize winner Tomas Tranströmer grasps this connection between wilderness and word in a particularly moving way in his poem "From March 1979." He describes it in his verses, but he does not resolve it:

> *Weary of all who come with words, words but no language*
> *I make my way to the snow-covered island.*
> *The untamed has no words.*
> *The unwritten pages spread out on every side!*
> *I come upon the tracks of deer in the snow.*
> *Language but no words.*[10]

For the American ecopsychologist Shierry Weber Nicholsen, "wild" is "a name for the way phenomena actualize themselves, the way they emerge from the fertile void."[11] The deepest criterion of the world's wildness is the discovery that this world organizes itself, that it brings forth from within itself, out of itself alone, without a drive toward a particular form, the longing for order, regular chemical reactions, and ultimately life, thereby creating yet more desire for order. Being wild means being uncontrollably alive, means unfolding oneself as part of a network of perpetual transformative relationships. Gary Snyder says: "Consciousness, mind, imagination and language are fundamentally wild. 'Wild' as in wild ecosystems—richly interconnected, interdependent, and incredibly complex. Diverse, ancient, and full of information."[12]

Erotic ecology—the transformative relational network of wildness— is the reality in which we exist and with which we mutually produce ourselves in every moment. "Matter" and "spirit," "nature" and "culture," "body machine" and "language" are not the mutually exclusive, foundational layers of reality to which a proper science must reduce everything in order to understand it. Fundamental reality is the creative wilderness in which everything interpenetrates, transforms, pushes itself into life, and carries death along with it.

This wilderness is neither spatial-material nor spiritual-abstract, but one *and* the other: materiality and nonmaterial meaning, life and decay, coherence and incompatibility. It knows no space and yet it fills all space. The "purely material" or the "purely spiritual" are its by-products, component parts of a *poetic space*, the inescapable wildness into which we are thrown by birth, even though we also already come from it. It is our illusion to believe that this wildness can be tamed—an illusion because our strongest instruments for taming it (language, imagination, tender contact) are themselves already wildly creative, imaginative acts.

Art alone extensively comprehends wildness without controlling it. Poetry alone realizes our aliveness without fixing a definition of it. This is why art is important to life—because it manages the uncontrollable within us without threatening to turn us into machines, without causing us to lose our aliveness. And the poetic is not only found in the reserve of "white cubes" in museums and biennials, or printed between the covers of books officially recognized as literature; rather, it is the core of every moment of creative reality, equally present in the creative babble of little children and in witty banter. Its surplus is the thin ground upon which reality is built, even if we do not notice it. It carries the flame of our life. We can experience this flame inwardly, but its warmth can also be felt directly on the skin—in the aliveness of other beings.

For the philosopher and musician Theodor W. Adorno, the role of art therefore was not to "imitate" nature (or the creation), which had been long considered the guiding principle of artistic creation in Western understanding. Poetic creation is not the copy of the world's creativity, but its central power. Art—at least, living art—does not imitate nature; it works like, or rather as, nature. It is an instrument and an embodiment of the wild. It transforms us by lending voice to the drive for transformation out of the "creative void."

The erotic of language: naming, transforming, creating

One of the central thinkers of such a universal world poetics was Johann Wolfgang von Goethe, the poet and natural researcher, polymath and actor in the self-directed amateur theater of a tiny court. Many times, Goethe described the transformative force latent in all of reality. Experiencing this force and its effects and then describing this experience is more precise than the exhaustive analysis of an ostensibly objective science. "Don't go looking for anything beyond phenomena; they are themselves what they teach, the doctrine,"[13] is Goethe's enigmatic, almost Zen-Buddhist attitude. They are themselves the doctrine because they have a transformative power when one surrenders oneself to them. The doctrine lies not in the neutral analysis of the world, but rather in entering into relationship with it and allowing oneself to be transformed by this relationship. Goethe called such an attitude "tender empiricism." In it, the self extends itself as an echo of a constantly vibrating creative potential and "every object properly perceived opens up a new organ of perception in us."[14] For a poet like Goethe, this organ appears especially in language.

Language is the ability to name things, processes, feelings, and individuals, thereby making them into something different. At that point, they are no longer objects, but rather imaginative fields of energy. The word given to each lends it a secondary materiality with its own sensory characteristics, which might be velvety, rough, hollow, or smooth. Things become voice and are not only transformed into signs, but actually receive a different body that they can call their own. Language doubles everything it names and thereby introduces an incompatibility into the world: It turns both sides—the object and its name—into something altogether oppositional.

Language allows the world to be named and thus possessed, but it also forces us to mistake the world and creatively transform it in order to compensate for this slippage. In speaking, I can transform objects and beings into my own voice and thereby assimilate them to myself, to an extent. I can roll them about in my mouth and let them bounce off my tongue. This enables me to release this newly created second body of something as its own vibration in space. I can transform things into something physical, into sound, making them tangible, and giving them the power to touch others, like a rapturous verse, which must be spoken in order to unfold its full power, which can seize the whole body and send a quiver across the skin.

In this way, language becomes an instrument by which one can multiply and also share this world, which belongs only to itself. I give out a transformed world and constantly receive back gifts of fragments of things in the form of voice, sounds, variations of touch, and embrace. I pass them along by adopting the transformative potential that lies at the foundation of language.

Language is a medium of metamorphosis between bodies and ideas and between different living bodies. In its potential to be a substance of transformation, the fabric of language corresponds to the nervous system of a living being. Language is thus another sensorium outside of the body, one to which we can help ourselves, one that has its own life, that calls itself forth and yet is also brought forth by me, that I organize and order, even as it also surrounds me.

The ecology of language thus forms another level of the creative ecology of mind in which we are embedded, another tier of a self-regulating and perceiving system that, in this act of perceiving, constantly brings forth new things. Language is wild like a landscape, and like a landscape it enables us to have an expanded experience of ourselves. Like the nervous system of our body, like the grass and the fireflies that are the nervous system of the landscape, language transforms everything it touches by transforming itself. Language touches what we are and makes it—and thereby us—into something else.

Language has its own logic, its own resonant vibrations, determined by the genius of every idiom's expressive possibilities, by its grammar, and by its history. Only because language exhibits this "unique vibration" can it respond to a stimulus. But it always answers with itself. And in this way, a sign system behaves in much the same manner as a life-form, whose sense organs always respond to the world with their "specific energy." An eye perceives everything as brightness and color, even a harsh impact against its glassy body. From contact between a resistant force and the retina, light emerges.

The quantum leap: a poetics of enlivened systems

Metamorphosis belongs among the foundational principles of an erotic ecology. In it, objects and individuals interpenetrate one another and transform within each other. It follows the rules of a connection that we simultaneously experience as a division, such that we must constantly compensate for it through new acts of interrelation. From this circumstance, we can derive further foundational rules for a creative cosmos.

1. Perception is always bodily touch (the contact of light particles on the rhodopsin deposits of the retina, the action of infrared waves on the feeling body, the movement of a membrane caused by contact with a scent particle).
2. Being touched means being in relationship: in the relationship that exists between oneself and another.
3. Every relationship entails a transformation of the two related poles.
4. Transformation is the translation of one through the medium of another (of the DNA code by the cell body, of the figures in a partner dance by the behavioral repertoire).
5. The act of translation by living systems can be described as cognition within a nervous system. Examples of such nervous systems include the brain, the immune system, the DNA switch genes, ecosystems, the planet Earth, codes of animal behavior, the vegetal rhizome, language, and other forms of expression (music, signs, architecture).
6. A nervous system allows for the perception of the other in the form of the self. It allows for the imagination of that which cannot be principally experienced as the self.
7. Every relationship is a nervous system.
8. A nervous system only functions because it is self-contradictory. Conversely, it can only be comprehended through a logic of contradiction.
9. This logic of contradiction is poetic logic. Nervous systems can only be comprehensively described according to poetic logic, although they can be neither defined nor designed by it.
10. Self-experience is the expression of one's own identity through the medium of the other. Perception of the other is the expression of the other through the medium of the self.

11. Self-expression means the simultaneously imaginative and real bundling of the world in a single individual or a single gesture.

12. This "whole in any little fragment" idea is the decisive moment for poetic experience.

13. The extent to which an experience will be effective (that is, the extent to which a change arises in a nervous system) depends on its imaginative potential—not on the material cause and effect, but on the symbolic significance of an encounter for the nexus of aliveness.

Vulnerability as a sensory organ

It is raining outside; it is still warm, but it is raining—I see it through the window. The dog doesn't notice and is resolutely determined to go outside. Once downstairs, I notice that I have forgotten the umbrella. At first the rain is cool on my skin. It has immediately soaked through my T-shirt. It strikes me coolly and then collects on my body in a layer of moisture. I endure it, standing in the rain for a moment and then slowly walking with my dog, who is without clothes and will be drenched, covered only by a porous coat of poodle fur. I let the rain saturate me, wetting my hair, the water running in small streams down my temples. I accept it without doing anything to prevent it; I accept it like an animal that belongs totally and completely to the world and is not separated from it, and suddenly I am filled with such joy and pleasure—in that moment, I know that nothing bad can happen anymore.

When the lightning begins, I move under the cover of a stranger's entryway opposite the Westend library. I wait as the rain rushes down and lightning flashes and flickers and unloads all of the electricity in the atmosphere. The rain drums down, the thunder rolls over the roofs, all of this desire for experience and being, this crashing manifestation of a sympathia universalis, which I am given to be a part of again today. I am in the midst of it—and at the same time, I feel what it is to be in the midst of it. I am the whole that experiences itself. I pursue theories, but I use the sensory eye of my skin to do so.

In this way, the world can regularly be experienced as a poetics of aliveness. This is a stirring affirmation, but one that breaks down every affirmative. It does not operate on a model of health, but rather one of creative fullness, or better yet: one in which the whole constantly learns about itself as an organ of perception, as an organ of transformative translation of the self into world and the other into one's own. At the same time, however, this translation forever misses the mark. I forge creative meaning only through the uncertainty of my self-assertion in the face of constant threat.

If one thinks in the midst of this precarious life, it becomes clear: Translation's perpetual missing the mark is, at the same time, its heart, the uncertainty out of which the self can experience itself every time through the other's response. Missing the mark is hitting it by missing it. Without this miss, the production of meaning would be impossible, just as life and perception would be unthinkable without a fundamental incompatibility between the feeling body and the abstract language of the DNA code.

The whole truth remains unutterable. Every creation is eternally incomplete. Meaning exists in every moment, but not a simple, unambiguous meaning that is true in every instance—rather a meaning of the moment in which a life-form produces itself according to its immediate needs to maintain health against a precarious fate. This meaning always carries its shadow along with it: the shadow that has something to do with the fact that the same action, in another moment, executed on a different life-form, might be deadly, misguided, unusable, calamitous.

To put it biologically: The incompleteness of creation is based in the incompatibility that lies at the heart of every creative system. Every cell that lives not only bears its negation—dead matter—within itself, but also is composed of that negation. This enables the cell (or the animal, or the person) to desire itself, to make its own being into a central concern, and this negation thereby constantly produces meaning—organized continued existence. Yet this continued existence does not resolve the contradiction. It abides it for a time. It transforms it into experience.

Into something inner, into something that does not inhere in matter, even though it completely depends on it. It transforms its own negation, sometimes, into the pearl of a moment of pure happiness. Nonetheless, this happiness bears its shadow within it, just as oyster pearls bear muddy grains of sand, the foreign body that had found its way into the tender insides of the mollusk.

The French-Algerian philosopher Jacques Derrida got to the heart of this connection in a short essay on poetry from the 1990s.[15] The text is also a statement on creative aliveness. What Derrida writes about the poem as the epitome of creative language could also be said of a life-form: It is never fully revealed, there is always a piece that is largely inaccessible because the bond to a living being, to a creative piece of reality, is always at the same time a separation. "No poem without accident," writes Derrida, "no poem that does not open itself like a wound, but no poem that is not also just as wounding. You will call poem a silent incantation, the aphonic wound that, of you, from you, I want to learn by heart. . . . The poem falls to me, benediction, coming of (or from) the other."[16]

Anyone who does not get involved with reality's "wild" shadow realm—and admit it in themselves—blocks out a significant portion of existence, if not the center. To be in the center is to be alive. To be alive is to surrender to the reality of a bodily existence that is never truly at peace and that gives birth to poetry and beauty out of this uncertainty. This requires the courage to no longer hide behind the illusion of immaculateness.

Uncertainty is therefore our more significant relational organ, the necessary precondition for our ability to establish bonds with other beings and with other people. Reciprocal attachment in which the self need not be hidden and in which the other has a chance to appear in the fullness of its reality. Every relationship is a collision that alters things irrevocably—and precisely this, this transformation of the self, signifies knowledge. As Derrida says, the accident alone, the touch as deformation, is what establishes a connection with reality. It is always both an experience of the world's weight *and* an experience of the self *and* the experience of the metamorphosis of the one through the other, open-endedly. Touch experienced through the other is the simultaneous experience of a self-reassurance and a wound.

To make this process possible, we must be prepared to take risks. Openness must always take on fear. Openness to the shadows—one's own and the world's. Openness means accepting the holes and tears without lamenting them. Having curiosity for what might come. Accepting that this world is a terrain of transformation, and that there is no transformation that does not also hurt. This pain is part of reality. Every experience is one of being wounded by reality, but also, unavoidably, one of being transformed by reality. And on the other hand, it transforms reality into something that is entirely one's own.

And so: acceptance. Accepting what is. The self-assurance of one's own penetrability. "Completely passive, all senses keened turned outwards," as the economist and humanist Manfred Max-Neef once described this attitude (and here again, another paradox). In a similar vein, the philosopher Krishnamurti writes that "observing without judging is the highest form of human intelligence," which makes possible all forms of transformative precarity and creation without attempting to rein it in or impose rules upon it, allowing it to exist uncontrolled and thus enlivening the painful desire within it.

Everything is there already

In the late summer, I walk along my linden alley after a long day of work. I inhale, almost in shock, the freshness of the evening air; I hear the grasshopper's innocent submersion in their song; I sense the stones' chill; I see the serene distance of the few stars already perceptible in the darkening sky; the mother-of-pearl steam mushrooms standing in the last light above the heat and power station in Spandau far to the northeast behind the long, shining arc of the train tracks. I am part of a silent fabric that embraces everything and carries it, never explaining anything—and never rescuing it either.

— *chapter five* —

EMBRACE

To love someone is to see him as God intended him.

MARINA ZWETAJEWA[1]

T he next day they drove to the Wannsee lake.

The next day. Blinding light. It is already 30 degrees Celsius (~ 85 degrees Fahrenheit), and it's still morning. As a child, you would put on shorts and a thin T-shirt and go out into the warm morning on your bike, a jam sandwich in hand.

Blinding city light. Berlin is not made for the mighty summer. It contracts. Staggered, it interrupts its own otherwise constant and all-too-evident activity. The city's soul snoozes in the sunshine. The streets stretch out like bluffs and tideways in a bay at low tide, and as on a beach uncovered by the sea, you can go walking at this time of day among things that are usually only half visible and hard to reach.

A morning that seems made to follow the night before. Unimaginable that a single detail could be different. Everything is true. Everything stretches out with warm softness. The presence of the summer lends all events an unwavering naturalness. In the morning, you see the other, in whose secure presence you have slept, the outlines filled again with form in the light. At night, we were enveloped by a sheet of darkness; we had the same rhythm to our breathing, and all that did not fit was hidden by the night, inaudible and mild.

On the way to the lake they take each other by the hand, boisterously skip over steps of gray concrete, greeting anew the gleaming day in front of the train station, then the woods. They hold their breath and touch the bright sandy soil with the soles of their bare feet. The skin has a sense of smell. It smells the ground and the pine needles and the dry grass. The skin is definitely the savviest sense—for a long time now it has gotten in good with the summer. They have become friends, for no reason at all, and dance with one another under the bright, leafy bows of the birches and pines. The body has become brothers with the ants, with the slender tree trunks. Fingers spontaneously reach for them in passing, touch their bark, a dry, calm, knowing nearness, the poise of the natural world.

Her skin is warm like the sand, her belly swells beneath the fabric, she is the queen of this Earth, but she doesn't know it, just as a blackbird does not know such things—but he knows it. And then he tries to forget it, to forget every word, to focus on the velvety ground of the woods beneath his feet, the velvet of her skin on the tips of his fingers.

That morning, there are the two of them, and there is the Earth they walk on. There is no one else. Not a single person from the packed city of Berlin crosses their path. There is only their path and the scent that wafts from it, the life that issues from it, a fine dust in the nose, on the bridge of the nose, on the lips (those wonderful lips), on the feet. As they walk, small atolls of fine dust forms on the tops of their feet. Everything around them is so unaware, so innocent and beautiful, simple and real and moving to tears.

And thus they step through the woods, speechless, lingering, dancelike, the dome of heat lining their path between the trees, and the rustling and crackling of nature's little arabesques accompanying them in secret tenderness. No questions, yet an ear wide open to the whispering of things, the gentleness of the skin that has found its counterpart. A preliminary festival, a segment of eternity, a promise that one can taste and smell—they stride through it in a short half hour, resigned and tipsy with happiness.

He holds her hand and she holds his, they hold each other's bodies tight, and those bodies touch the plants, they look into each other's eyes, and those eyes reflect the glistening light of the summer and are filled with it until they overflow and bubble over onto one another. They hold each other by the hand because nature holds them; being surrounds them because it is being, not becoming and fading away—life is on their side. It leaves them nothing to doubt, and they joyfully cast it aside, that doubt; distrust has been cast long ago into the trash on the side of the road. They gave it up at the beginning, and now they are completely naked. Nobody can see them that way, naked as the birds, naked as the blades of grass. They walk with one another and breathe the air that is so favorably disposed to them.

The two of them are completely in the world, and they are completely themselves. They are individuals within a whole, and through them, the whole is able to show its full majesty. They know that it is only a moment, a blink of the eye—the few minutes that they walk on the springy grass of the Grunewald through the air scented with pine resin and the perfume of a dry forest floor. They know that everything will pass again, and this fact alone allows them to grasp the enormous intensity of this moment. It is fragile and it is comprehensive. Intoxicated, they sense their uniqueness, but they sense it only because they are part of something enormously dense, unbelievably complex. They part the naked sand with sensitive toes, they touch the other's skin, made breathless by the ever-new and long-familiar surprise. They are connected and divided at once, inescapably.

The world—sucked into the lungs, suffered as explosions of light on the retina, felt as vibrations on the surface of the body, completed by the echo of the beating heart, shot through with birches and pines, with impatiens and blackberries, sung through by chaffinches and crossbills, tortuously, surprisingly, unfathomably, and overpoweringly filled with happiness—reveals itself in these short minutes as a beautiful complication.

That day long ago during the summer on the way to the Wannsee lake was a scene of relatedness and its incomprehensible rules. Everything lay there, but nothing was known. That moment in which a person experiences his or her existence intensively, with his senses and with his emotions, contains the whole spectrum of criteria for being alive in a creative world. Experiencing it requires no special experimental arrangements, no special aptitudes, and in truth, no mediation, even. We already have everything that we need, because we are always part of the ecosystem that we participate in and seek to understand. In our bodies, we are part of this world. Everything is there to comprehend the mystery of existence, to experience this world in its erotic communality of boundaries and exchanges. Our existence as a feeling body outfits us with all of the instrumentation we need to not only live our lives, but also to comprehend them. The body is the most sensitive instrument imaginable for providing information about life.

In every intensive moment of experience, we witness ourselves in relationship. We experience the fact that our own emotions are created in a particular environment, even as that environment is also being created. Through the experience of light, for example, our skin transforms invisible waves of energy into the self-perceptions, "pleasant and useful (meaning: bright)," or "destructive (meaning: blinding)." At the center of existence is the experience of attachments that have existential significance. Relationship is the foundation of our being. And relationship is the foundation of our social reality. We live as relationship.

This is why this section is dedicated to the connections between people. These, too, are guided by the same universal principles of aliveness that characterize ecosystems. The erotic and love are instances of the same creative ecology that guides our metabolism, our perception, our imagination—according to whose principles, for example, a bacteria cell freely and creatively generates meaning as a "surplus" of its existential encounters.[2]

From the beginning, we are ourselves

All connection begins with childhood. A person is the result of an encounter between a sperm cell and an egg cell that together transform into an embryo. This process has carried on for generations, indeed for millions of years, ever since the time that higher animals engaging in sexual reproduction sprouted from other life-forms that reproduced by dividing their cells. The history of relationships reaches back to the moment that life emerged. Every one of us is a survivor from the very first inception of being. In this sense, we are truly seasoned practitioners of successful relationships.

In the mother's body, our connection unfolds in an illuminating way. A fetus is an independent being, yet it is completely reliant on the mother for sustenance. In the placenta, nourishment enters its blood from hers, but the two circulatory systems remain separate. The two can even have different blood types, and if the systems mixed, in this case it would lead to clotting and to fatal shock. Mother and child each have their own bodily and soul identities, but at the same time they are engaged in an unavoidable exchange that lies outside of their control. As such, the mother's feelings drift into the embryo by way of messenger substances that move through the placenta. The embryo becomes flooded by strange emotional conditions, just as a hiker along the coast might encounter the power of a storm. Nevertheless, the child is an independent being, not an outgrowth of the maternal body.

Herein lies the source of a misunderstanding that embryologists and childhood researchers are only beginning to clear up now.[3] For a long time, they assumed that mother and child formed a "symbiotic union," that the child experienced the mother as a part of itself, as the extension of its own body. Many psychologists assumed that even after birth, outside of the mother's body, the infant lived within this "primary symbiosis" until it was two years old. Mother and child were not separate from one another in the newborn's experience. The child must rather learn this division during the first years of its life. Many psychological concepts are based in this assumption. Many researchers assume, on the basis of this idea, that an unavoidable trauma is our common fate in the cradle of early childhood: the shattering of the perfect union with the mother.

But this perfect union never existed. The situation in the uterus is not one of bodily identity. Every pregnant woman will confirm this. At the very latest, when the child moves in her uterus, when it begins its own life at 4:30 in the morning, it is very clear that the mother is carrying another person within herself. In the uterus, the fetus does indeed exist in a form of intense relatedness, but it is, at the same time, its own individual. The fetus has a clear, unique identity and experiences this as such, and yet this identity must also be constantly created through interplay with an other. This first other is the mother: She provides her child with oxygen and nourishment, she cleans its blood, she rocks it in amniotic fluid during sleep. She gives the embryo identity by tending to it with her own body. We see here a creative exchange through which one subject enters into the world and another is irrevocably changed.

Our phase of life in the mother's body is the ecological archetype for a connection-in-separation, and thereby the model for what the child will later try to create outside the mother's body. The healthy, normal condition is one of close connection, but not symbiosis. Only because the mother is herself alone, and not her fetus, is she able to give the child what it needs. A connection is possible only because both are distinct from one another.

Rather than speak of a primary symbiosis, we can therefore speak of a primary separation: The baby experiences the mother not as a part of itself, but as an other. It experiences her as a nourishing landscape that establishes the conditions in which it thrives, without understanding why. Floating in amniotic fluid is a passive state of being, even if it also provides everything needed: a suffering redeemed by the liberation of one's own subjectness through birth. The newborn is indeed less protected, but it has other means to control its surroundings. After birth, it must be seen whether the world will continue to support the newborn, whether something more than metabolic necessity will allow it to flourish; namely, the conscious wish of another human being to keep it alive. The trust that a baby can form toward its mother during this period of time becomes the measure of its trust in the world. The strength of this first trust continues to determine its life long after it has grown up.

After birth, the mother's role is to transfer the formerly unconscious trust in the world that the child experienced in the mother's body. She has to gradually convince the child that she alone is not the world. And she has to give it the chance to experience that the world is not the only thing that determines the conditions for the child, but that the child itself has a creative power over how things turn out. She must love the child away from herself. In the course of childhood, the child should unlearn that another person determines everything, godlike. It must comprehend that the mother is an individual, and that it too is an individual, and that with its own powers it can exercise productive influence on the world.

An infant knows it is an I

A child is able to form its own identity because it experiences an other that displays independent, individual, and uncontrollable emotions—in other words, an other who is a subject—and because this subject enters into a well-meaning relationship with it. The clear experience of the boundary, the encounter with a You, is necessary for the development of the I. Groundbreaking research on this topic has been done by the American evolutionary psychologists Andrew Meltzoff and Keith Moore. They have shown that even newborns react to a person's facial expressions—specifically, by imitating them. When a test subject smiles at a newborn infant, the child gently curls the corners of its mouth upward. If the test subject purses his or her mouth, the baby also tightens his or her lips.[4]

At first, the researchers could hardly understand this ability on the part of the newborn. They were biased by a picture of the infant as a passive symbiosis machine and hardly dared to imagine that a newborn would have experiences of itself comparable to those of other people. But the infant's ability to imitate left no room for doubt: Immediately after birth—and thus also in the mother's body, from which the infant had only just emerged—a child knows that its inner experiences are connected with its body and that they express themselves there.

This means that babies never have to learn that their experiences belong to themselves. And it is also clear to them that others can read these experiences on their surface. All of this knowledge arises on its own from the logic of a vulnerable body that gains an inner perspective through its relationships to others. It is our primary, genuine experience. We know what it is to be a subject that feels something "inwardly" and shows these feelings outwardly. The child knows. It does not have to learn that it is an inside with an outside. It reacts to the presence of a communicative outside with a corresponding inner gesture. When an infant imitates another's smile by smiling itself, and experiences the accompanying feeling of joy and sympathy, it understands how the other person feels. It is in a position to do so because it itself is the connection between outside (the smiling face) and inside (the feeling of released joy).

The parents' role in this relationship is to strengthen the child's sense of security and not to hinder its opening up to the world. Of course, this only works when the people in the relationship can express their feelings. A mother who smiles but feels no joy in doing so shatters the correspondence between inside and outside. If the other does not display emotions because his or her own emotions are disrupted, then the child will unlearn the connection between outside and inside, and its own affect will become a mystery to it. For being able to love means trusting in the connection between inside and outside.

And so again we have a paradoxical situation: An infant knows that it is an independent I with its own distinct body, but at the same time it must practice being this I fully and completely. Here again we encounter that paradox, "Become such as you are," expressed by the Greek poet Pindar, which characterizes erotic ecology so deeply. The core of identity must take root and sprout branches through exchange with its surroundings in a mutual play of transformation. The main pursuit of all human beings during the first years of life is to perceive and unfold the fascinating groundswell of their own identities in exchange with the world. This is also the ideal of a successful partnership between adults.

The psychologist David Schnarch states that "the fundamental link between disengagement and connection is even evident between infants and mothers."[5] With this, he, like other researchers, gives up on the idea of a "primary symbiosis," and instead emphasizes what could be called a "primary dialectic"—a fundamental separation in connection. His colleague Shierry Weber Nicholsen affirms: "This capacity [to be alone] depends [. . .] on the experience of being both separate and merged with another. [. . .] We are first alone in the presence of another, the mother, and it is this experience from which we develop the capacity for solitude."[6]

Researchers observe that from the very beginning, babies have their own independent emotional life. They sometimes enjoy being left alone and are capable of occupying themselves from early on. Infants break eye contact with their mothers when they want to be alone. This also shows that from the beginning, people are not outgrowths of an identity shared with their caretakers, but rather beings with their own, clear feelings.

We have the inner experience of an independent identity even within the mother's body. At the same time, in order to grow, this identity must be recognized, must be given its own place, and must find a fertile environment outfitted with joy and generosity in which it might unfold itself. The important thing is the interplay between the offering and the recipient of that offering, as is the case in cellular self-production out of soma and DNA, in the metabolic process, and in every experience. It too is a question of both sides, not just one of the two. Overlooking this fact is the source of a decades-long misunderstanding, exemplified in the nature ("genetic data") versus nurture question. The two sides exist independently, and at the same time, they mutually cocreate one another.

The face that a child gazes on with the feeling of kind affirmation is the figure of love, is love itself in a bodily form, benevolent aliveness that gladly gives in order to awaken more of itself in the world. It thereby stimulates within the child the unconscious wish to pass along this affirmation.

The child smiles back.

The other half was never lost

Smiling is one of the most important forms of exchange between human subjects. The eyes aglow, they say: I see that you see me, am glad to be seen by you, and want you to see this. The I is affirmed through the other. It neither merges with it nor stands completely apart and alone. It sees the other and is seen and exists in a precarious balance with it. The I constantly relies on the fact that it is perceived by the other, and that the other signals to it: You may live and be enlivened. The other thus has but one thing to do: to admit another's aliveness, to allow this precarious balance. It has no particular resources to give that are essential to life. It loves by giving to the subject it encounters that which we all have to share in abundance: a climate that desires life.

From the first people you relate to in the earliest stages of childhood, to later partnerships, to the relationship with your own children, the essential role of the other does not consist in furnishing your own I with useful things. Of course, a significant portion of the Western understanding of romance is based precisely on this perspective. Here, the beloved always appears to be in possession of something that I lack—a certain beauty, a strength, a resource, or a gaze that sees something in me that I do indeed possess, but cannot recognize.

The craving for love, according to one cliché, is based on the desire for this something that one cannot produce alone. "I love you because I need you," it says. Or rather: "I do not love you anymore, because I no longer need you." The philosopher Plato recounted the saga that we are all searching for our "lost half." In Plato's famous *Symposium* dialog, the priestess Diotima tells this story as a divine myth. In the beginning, all people were complete and whole like perfect spheres. But then they were split into two halves. From then on, each of us is looking for the missing piece that will allow us to become whole again.

Since then, we tend to understand love as something wherein the beloved provides me with a capacity that I have long been lacking. Indeed, we could perhaps say that the role of love in Western culture has increasingly taken on the responsibility for balancing out a deficiency in a person's own I. The philosopher and theologian Christoph Quarch says: "If I fall in love, then it will be with a person who I suspect can give me something that I lack: that she can satisfy my wholeness."[7] In this way, we live in a Platonic age. The Parisian philosopher and artist Fabrice Midal, who has a very different perspective on the erotic, writes: "But the genius of love is actually that the other person who loves me sees me as I am, whereas I am unable to do that. He sees the utmost in me."[8]

Living reciprocity is an exchange, not an act of providing someone with resources. It is an interplay between two identities from which both emerge transformed. Each partner in the connection is unavoidably alone and a self—just as the fetus is alone and a self. But this being as a self exists in constant dialog with the other. This enables me to transform, just as I enable the other to transform. The creativity of this transformation and its ability to make life possible determine the extent to which both partners in a relationship can "be themselves." The other makes a space for me in reality. We unhappy lovers in our hedonistic civilization have a tremendous amount of relearning to do.

The role of a partner in a successful relationship is to increase our collective aliveness *together with me*, thereby increasing both his or her own, as well as mine. This is true of love, of friendship, and of parenting. It is also true of the connection with animals, with plants in my garden, with the biosphere as a whole. Every successful relationship is ecological: It productively integrates into the living network with the goal of supporting aliveness as such and aiding in its growth. Love is thereby a practice of relatedness that makes the whole more real. It makes it more real, not nicer. For both light and shadow, pain suffered by both the self and others, are part of this reality.

The age-old myth recounted by the priestess Diotima in Plato's *Symposium* is thus already overshadowed by a cultural deficit: Beings that seek their second half exhibit the bitter privation of the needy who are not allowed to be themselves. The wise Diotima articulates a relational-ecological dilemma. She is expressing the condition of painful separation as a normal state of being in the world, but it is the outcome of a failure to relate. There is a reason for this failure. For Plato has her speak according to the spirit of separation that pervades his entire philosophy and that has exercised an incalculable influence on Western thought.

Diotima's stance follows the general program of Platonism: The world of bodies—in other words, the world in which we live—is already the result of a separation. Reality is isolated from actual things, from the "ideas." According to this thought, to live is to strive for a completeness that cannot be found here, but can be found only in a transcendent-ideal world. In Christianity, God was placed in this position of ideal truth, and "actual" life came only after death. Modern science filled the Platonic ideal with its goal of eradicating all earthly evil through analytical understanding and technological improvement. In the current marketplace of love, the ideal has again been placed at the center of our desires. All of this has a common denominator: The real world is viewed as flawed. In love, a piece of the ideal world beyond is supposed to become real, but the result of this is an act of controlling, not one of allowing the other to be enlivened.

Plato, and many thinkers of the following millennia who were influenced by him, takes for granted that the human I can participate in the world of the ideals. The body is the only thing that imprisons us in the dark cave of earthly life. In the history of philosophy, this notion is called the "radical transcendence of the ego." The I actually does not exist in this earthly world, suffering and loving in the experiences of the sensory body—rather, it exists somehow as an otherworldly abstract principle of knowledge. It is consciousness, spirit, rationality—and therefore structurally unrelated to the impenetrable drives here below.

For a philosopher like Immanuel Kant, who made this theme into the driving force of his life and thought, it was an inexplicable mystery how this cut off, "otherworldly" I that was imprisoned within the body could ever obtain knowledge of the world. In this worldview, love—the striving for completeness and knowledge—has little to do with the principles of aliveness and its living relationships.

Biological anti-platonism

If we go back to the origin of our experience, to the situation of the infant in the mother's body and after birth, we find that the I is not cut off. But it is also not predetermined or produced by its exterior. It is both: radically bound to its own body but also implicated directly in the ecological whole by this body, which is a sensitive organ of exchange between other such bodies in a living world.

One's own I is, philosophically speaking, *radically immanent*, wholly and completely a feeling part of the world. It lacks nothing because it is not cut off, but always already in relationship. It is not a half, but a whole, and at the same time, it is completely dependent on exchange with others. Because it is immanent, dispersed in the world, it does not have to find resources that an other might deprive it of. Its concern is rather to make this immanence completely visible, which means begetting as much aliveness as possible.

In my eyes, the philosophers' "radical transcendence of the ego" reflects a pathological condition of the soul, a trauma. It results from the pain of having one's own aliveness trampled and disallowed. Most of us have experienced this. Such a trauma propagates itself. If the I is irrecoverably shut up in itself because it was not seen by a benevolent gaze, then it will turn everything with which it comes into contact into a means of alleviating the torment of this isolation. Every relationship partner turns into a resource for survival.

It is very interesting to see how such disruptions of our natural capacities for relationships can influence an entire world. They cement the separation but also generate an obsession to overcome this separation, no matter the cost. Our current fixation on success, constant competition and fighting, and war as the father of all things feeds on this obsession: It is the drive toward total victory that alone promises to overcome the solitude that is so deeply felt.

We also find this understanding of attachments in evolutionary psychology. This discipline, which tries to explain the features of the human soul in Darwinian terms, cultivates a genetic Platonism. Darwinism begins with the idea that every being is a bellicose ego in matters of its own genetic frequency—and that the world is only truly real to the extent that it contributes to the success of one's reproduction. Mainstream Darwinism considers partner selection and child-rearing solely according to one perspective: How will this action result in an increase in the number of my descendants? Accordingly, it understands everything as an investment. This understanding dominates our unconscious imagination of the world and ultimately leads to the complete elimination of the self. If the self has nothing to do with the world of bodies and if the world is considered only according to an economy of survival, it seems only logical to brush the self aside as a partially useful illusion.

Is it not striking to recognize one of the greatest driving forces of our current civilization in this basic Platonic attitude, which has accompanied Western thinking for millennia? This driving force is in the way we organize how we live together, but also acts in every one of us. It has taken the form of a persistent need to overcome an alleged separation, whether by seeking a partner in order to balance out our own deficits, by expecting science to illuminate the shadowy realms of reality and provide for unambiguous order through knowledge of ideal relationships and conditions, or by sacrificing aliveness in favor of an allegedly brighter economic future.

The Western concept of love since Plato seems to fall into error in many ways. It thinks of love as a deficit: a longing for something we do not yet own. It is a concept born from the thinking of possession, from the view of the ego as needing to have, rather than as needing to be. This thinking therefore desperately clings to the idea that choosing the right partner will result in a wholeness that cannot, however, exist in any creation. For millennia, this Platonic tradition has tried to overcome the experience of separation through the right actions, through technology, self-optimization, the subjugation of others. But separation is the central marker of creative reality. It cannot be overcome, only *transformed*, if we do not want to destroy this reality.

We should therefore understand love as a practice of relatedness between two poles that cannot be united with another. It is that which unfolds through creativity and transformation and thereby makes the whole more real. This reality will always include light and shadow, joy, but also pain experienced by both the self and others. The philosopher and mystic Richard Rohr writes: "The true contemplative mind, however, does not deny the utter 'facticity' of the outer world. In fact, much of its suffering comes from seeing and accepting things exactly as they are."[9]

If such an attitude is not to become insipid and mawkish, then the challenge of it is this: to decide in favor of aliveness is to be ready to fight for it at any time. Love is thus no longer the desirous search for that which I do not have, but rather a clear feeling that I already have something—namely, the aliveness that must be defended. It empowers one to fight for the right to become alive in the practice of one's own love, and thereby to enliven others. At the same time, it does not exclude the knowledge that nothing can alter the fundamentally tragic coordinates of reality.

Love: our ecological model

True order is therefore always ambiguous. In many ways, we experience it as lack. At the same time, we already have everything that we will ever need. We have simply hidden it from ourselves. "We must learn to accept paradoxes, or we will never love anything or see it correctly," says Richard Rohr.[10] Our living identity—itself ever contradictory—enables us to have this perspective. Our capacity for double vision is already present.

And if it is not, because it was traumatically destroyed in childhood, then no later relationship partner will ever be in a position to repair it. Only I can do that for myself.

Identity must produce itself. Every organism is the temporary result of an endless negotiation of identity between the billions of cellular selves that make up its body, between its body and the environment, between its imaginations of itself and the echo that others reflect back to it. The identity that lies at the foundation of one's own self can only come forth lastingly and productively when this self-production occurs in acts of mutual transformation. Like the coherent structure of the cell in the metabolic process, it must constantly be created and affirmed through exchange with other identities. Otherwise, it will be starved out, so to speak; it will begin to crumble and to develop fantasies that only provisionally accord with reality.

Here again, it is a matter of walking a fine line between one's own material self-sufficiency and an overemphasis on the other and its influence. Within every being, identity forms through autonomous relational processes and must, at the same time, be constantly affirmed and reflected from without. Both sides are fundamental, and they depend on one another. I myself am alive. The other lives through my aliveness and respects it, as I do its. These alone are the ingredients for a successful, humane ecology. Like the condition of the fetus in the mother's body, it is a connection-in-separation. And precisely because it is both, because the results are always open and uncertain, it is so crucial to develop an etiquette of the bond wherein both sides make an effort always to put aliveness first. The love of life counts for more than the obsession with a particular person.

For too long we have thought about love without considering our ecological reality, our reality as parts of unfolding relationships. We have learned to describe love as an irrational feeling reserved for human beings alone—or we define it as a chemically controlled, evolutionarily necessary survival function, as has become increasingly common in recent years.

But love—the experience of a deeply meaningful relationship between people—contains two things at once: action and inner awareness. It is the experience both of living out one's existence in flesh and blood, and of understanding this existence. In the feeling of love, we inwardly comprehend the practice of being alive, which is manifested through the body in contact with other bodies. This feeling is the strongest connecting link between the emotional interior and the bodily exterior, solid proof that both are aspects of the embodied existence.

In this, love becomes a practice of realizing existence. This existence is always a matter of calling oneself forth, together with the other, and thereby constantly transforming both oneself and the other. We can understand reality as a network of relationships that are relentlessly reforming, thereby becoming a perceptual system in which "the whole" regularly reexperiences itself through the individuals that compose it.

This is the essence of the erotic: All-encompassing and inescapable relatedness makes the world capable of perception, capable of learning, imaginative, and sensory. In the erotic encounter, the world touches itself and thereby arrives at new knowledge of itself. Relationships are its opportunities to have experiences. The way in which we connect with one another expresses a particular variant of the unfathomably complex relatedness that lies at the foundation of all things. There are some things that only occur between people, but there are also many aspects of this relatedness that are universal to all species, and some aspects compose all that is made of matter.

Our civilization currently understands very little about the principles of such life-inducing bonds, which have shaped the universe from the beginning of time. *Homo sapiens* regularly tears the fabric of reciprocity that composes reality, and again and again severs these bonds. Our relationship crisis manifests in numerous ways, such as the disappearance of species and habitats, the ordinariness of failed marriages, or the constant increase in cases of depression and personality disorders.

How we love is nothing less than how we tend to our ecological attachments. Because love is the site where inner and outer dimensions meet and mutually comprehend one another, there is an ineluctable feedback loop between the way we treat the world and the depths of our love.

We should recognize that we are inescapably woven into a relational fabric. In the Platonic tradition, in which our I belongs to a sphere of the ideal and of unlimited freedom, such an insight sounds a lot like determinism. But following the principles of reality does not mean being unfree and determined. Rather, it demands the recognition that there are fundamental conditions of our existence. They influence us, but we can influence them as well. Natalie Knapp understands this as follows: "We are molded by chance, playful encounters with nature and at the same time hold the thread of our own history in our hands. Both facts together shape our lives."[11] The reality, for which we have so little respect currently, is not reducible: it is unpredictably creative. But even so, it follows certain rules, just as the butterfly's flight through the air is unpredictable and also bound to follow physical laws. The unpredictable flight of the butterfly is indeed only possible because it lawfully applies the principles of physics.

Relationship as the epitome of reality

The philosopher Fabrice Midal says: "Love is not one ancillary feeling among many, but rather the dimension in which our existence materializes."[12] With this, Midal looks at things the other way around: Life in the fullest sense is relatedness and thus always already a practice of connection. All actions, all connections, and all experiences are only understandable and meaningful when they are fulfilled as an intensification of our relationships, and thereby of ourselves. On the other hand, distressed relationships are ones in which the foundational principles of erotic exchange that I am investigating in this book are violated or flouted.

The erotic that we live out represents a small section of the erotic whole, but one in which its collective fullness is representatively active. It is a facet that reflects and refracts the whole, one that we can participate in for a short sequence of moments because it embraces and seizes us. It captures us when the rules of creative reciprocity are allowed to take effect. The experience that individuality can only be asserted through creative reciprocity can inspire us to form more enlivened human bonds.

The rules of erotic exchange are open in regard to the outcome, but they move strictly according to the principles of the creative imagination. These principles describe the occurrence of something new through the connection and transformation of two irreconcilable poles. And it is here, precisely, that these principles meet up with the reality of human relationships. Reality is contradictory, and all behavior can only be decided or evaluated in concrete situations—forecasting is impossible. A first approximation of the rules of erotic exchange might go somewhat like this:

1. Every participant can produce his or her aliveness only by him- or herself and cannot acquire it through another.
2. In order to experience ourselves as alive, we must bestow aliveness on others. The You precedes the I.
3. The other's offering is a gift, not a reward for a service.
4. Through giving back this gift—the other who sees me in my reality—I can become what I am.
5. A relationship is successful when it increases the aliveness of all.
6. A relationship is successful when all can reveal their needs.
7. Any relationship is imperfect. Its imperfection alone is what allows it to continue developing.
8. A relationship is unity in separation.
9. Each partner is the owner of his or her own death.
10. A relationship is play.
11. A relationship constantly transforms both partners such that each perceives and comes to know him- or herself through an aspect of the other.
12. A relationship is a nervous system.

In our human bonds, it's all about the whole, whether we own up to this or not. It is about how we can unfold ourselves as a creative part of reality's contradictory and paradoxical fabric, and also about whether we grant this to others. It is about how much we can accept, in our core, the productive possibilities contained in these living cosmos; in other words: It is about how alive we are able to be.

The couples' therapist David Schnarch identifies the basic problem of aliveness as a "two-choice dilemma." By this, he means the contradiction in life between fusing with the whole and individuating, the ecology of incompatibility inherent in life itself. After many years of practical work, Schnarch writes without illusion: "Our problem is not the two-choice dilemma itself but our refusal to face it, our unwillingness to meet life on its own terms."[13]

Relationships, regardless of which ones from the dense web of reality, are always a mediation between nothingness and totality. The fundamental problem of every human relationship accordingly revolves around *nearness* and *distance*, or (in a slightly modified formulation) the tension between *self* and *other*. How much can I remain "I myself" in a relationship? How much of myself must I give up for the benefit of the other? Where do I draw the line? Is there even a Me there before a You has begun to truly see me?

For the influential American psychologist and author Ernest Becker, life is characterized by a basic "ontological" or "creaturely" tragedy. It is the clash between two basically irreconcilable needs—one for connection and one for autonomy. We have seen that the central tension of all human relationships also results from these two poles. For Becker, we are living now in an epoch that believes it can resolve this tension in loving relationships between two partners: through the romantic union with the *right* partner. In the "romantic solution," we fix our "urge to cosmic heroism onto another person in the form of a love-object."[14] Romantic love, and the ensuing connections, confusions, separations, the experiences of ecstasy and hatred that come with this, all follow from our attempt to escape death.

But exactly this seeming solution is bound to fail in a painful way. Becker writes: "We enter symbiotic relationships in order to get the security we need, in order to get relief from our anxieties, our aloneness and helplessness; but these relationships also bind us, they enslave us even further because they support the lie we have fashioned." The object of love, according to Becker, becomes a god—and can therefore transform just as easily into the devil.[15] For not the character of the other, but "life itself is the insurmountable problem."[16]

Life's tensions cannot, as a matter of principle, be resolved into something free of contradictions, because they follow the basic tension of all creation. They are modes of appearance of the tension that produces all things—that incompatibility between infinity and concrete form, between genetic information and the body involved, between inner experience and outer matter. Like these other tensions, the contradictions of life cannot be cleared up. Rather, the creative answer to living reality is to bring its duality into a dynamic balance.

Seeking this balance—and constantly failing to achieve it—is the central drama of human connections. It is also a phenomenon of the ecology of reality. How we enter into relationships depends significantly on what we consider reality. At the moment, we regard it always according to the optics of salvation—the self-assurance, dating back to the early days of modernity, that we can handle God's affairs better than he can.

But what if actual salvation did not lie in straightening out the difficulties of existence (my house, my car, my husband, my children, my kids' school graduations, my Labrador retriever) but in abiding and transforming them without pushing them away? Admitting to ourselves that reality cannot be escaped, and that our greatest creative potential lies in precisely that fact? Receiving the call that reality cannot be escaped, but remembering that it can be caressed and helped?

Fully I, fully you: the magic of skin

One of the greatest pioneers of radical immanence, a deep feeling of kinship with the world and a practice of love as loyalty to its beauty was the French poet and philosopher Albert Camus. Camus had suffered bitterly during the rationalist 1950s because he had founded a philosophy based in the sensory feelings of the body. He was largely excluded from intellectual circles by his colleague Jean-Paul Sartre because Camus's intuitive, sensory, and poetic manner of thinking did not suit Sartre. Plato against Pindar. Nowadays, Sartre has largely become a part of intellectual history, but Camus is only just being rediscovered.

Camus's instrument for connecting with reality was the skin, which opens itself up to the lavish southern luminescence. Born in Algiers under poor conditions, the later philosopher had in childhood only the munificence of the warmth and the beauty of the light and the sea, the sensory encounter with the warm and clement Mediterranean world. The young Albert spent all of his free time on the beach with his friends, caressed by the sun, bathed in its blaze. Having all of his senses attuned to light rescued the child from many deprivations, particularly the mercilessness of his cold-hearted grandmother. "In the light, the earth remains our first and our last love," the writer would later say.[17] And then he reiterated: "The world is beautiful, and outside it there is no salvation."[18] That was a philosophical experience. But the instrument of this experience was not "pure mind," not the intellect, but the body, the wisdom of the skin.

The skin, that site of clement consciousness in warm light, in the cool ocean, on the hot sand, is always a medium of both connection and separation. Camus's contemporary, the philosopher and researcher of perception, Maurice Merleau-Ponty, observed this as well. Merleau-Ponty was fascinated by the fact that every time I touch someone or something, I am simultaneously touched by it. This is what results in the partners of a loving relationship experiencing such great delight: If I stroke your skin, I am feeling the surface at which you begin and through which you impart yourself to me, with a gentle shudder, an involuntary approach, or a delicate withdrawal. You feel me, the tips of my fingers, tracing over your surface. And my fingertips feel you. But I can only feel you because I am also feeling myself: The touch of my skin signals to me the touch of your skin. You have to touch me in order for me to touch you. And you only feel me because you are also feeling yourself at the spot where I am exploring your boundary with my fingertips. Only by feeling myself can I feel you. And only by feeling you do I feel myself. To be sure, I exist before I have touched you: I am already a subject with a body. But I am conscious of this body to a much greater extent insofar as I can allow it to be touched by your body and thereby feel myself. Both touches are inseparable from one another. They are bodily, materially, physically inseparable and can only occur simultaneously. By touching you, I must open myself up to be touched by you. In order to be able to open myself up, I must be able to open myself to someone: I must allow a concrete other to touch my boundary.

"Chiasm," the "crossing of the world," is what Merleau-Ponty calls this foundational dimension of our perception of both self and other. For him, it follows that all experiences in reality can only be described as reciprocal, as a communication web made up of the experiences of countless distinct individuals. He calls this living, interwoven reality of collective experiences the "flesh of the world." With this, the thinker gets to the heart of bodily, creative aliveness, that aliveness that is poetry inside and a breathing body outside. Flesh of the world like the rising skin of young life, like rosy cheeks, like bursting apple buds, like supple sea anemones rising with the swelling water, like the light that falls from the blue of the sky—but also flesh like the sound of a verse, like the inner sensation of a feeling. The flesh of the world is not merely fleshly. It is not the biosphere but its reciprocity in an endlessly creative relationship. Merleau-Ponty thought that everything we perceive is connected with us such that it also perceives us. The tree experiences its presence when we look at it. The snow that slowly falls perceives our presence when we leave our footprints behind in it. The Earth perceives the moon's presence and bends itself toward it with its waters.

Because I can feel the world with my skin, I can also feel myself; and because I can feel myself, I can feel the world. In the mutual touch of the skin, both are there at once, born together: I, subject, individual—and the world that surrounds me and of which I am a part. Reality actually only exists in this moment of touch from which both sides emerge; it exists only in the sensitivity, in the tangibility, and in the attention with which I perceive both myself and this sensibility to other at once. Does this not open up a fundamental, erotic connection to reality? The world feels me like a lover. The world is tenderly prepared to receive me at any moment. "The ability to be perceived is a basic form of attention," wrote the poet and philosopher Novalis, fully in the spirit of Merleau-Ponty, but 150 years earlier.[19] Novalis saw the whole world as flesh reacting to a self, simply through its potential to become a partner in the exchange of perceptions that is the flesh of the world.

This close connection between self and other can be sensed in all dimensions of our organic existence. Our own body's immune system, for example, can only recognize an antigen when it is "presented" on the surface of one of its own cells. The immune cells identify one another as component parts of the bodily self and only therefore are able to distinguish an other, which they can recognize only *as a modification of the self.* And the antigen smuggled into the body, the parasitic cell, the virus, likewise perceives the immune system cells to some extent, because it is truly seized, processed and anchored in the surface of a cell by them.[20]

Let me see your eyes: the "extended vision"

Our gaze also exhibits this reciprocity. How is it possible that we can discover ourselves in the eyes of another person, insofar as we discover the other person there as well? This is and always has been a mystery. The glistening pupils of the other do not reflect us alone, but us in connection with the other, the person we are beholding and who looks back at us. We see how we are seen and allow ourselves to be observed.

A friend of mine with particularly beautiful eyes, warm and expectant as a fresh summer night, maintained a long-distance relationship with a partner for a long time. She told me that often, when she spoke with her boyfriend using a chat program on the computer, she would move her eyes so close to the onscreen camera that her lover's monitor would be filled solely with the sight of her gaze—that radiance that perceived the other's shadow and returned them, transformed by its own feelings. My friend had turned it into a habit, activating the interconnection of a reciprocal gaze in this way, even across time and space, and for the two of them it became a ritual over the years. "Do you want to see my eyes?" she would sometimes ask him, from a distance. This works even over Skype. I have tried it. It was as though the distance between us was obliterated.

Our eyes not only take in light but also radiate it outward in a mysterious way, as though they were heavenly bodies. Eyes are the bodily organs that constantly attest to the fact that we experience the world by transforming it, because they express this transformation. They see that I see that they see. The French poet Paul Valery thought: "You take my appearance, my image, and I take yours. You are not I, since you see me and I don't see myself. What is missing for me is this 'I' whom you can see. And what *you* miss is the 'You' I can see."[21]

I have to look at someone in order to be seen by him; I make myself visible through the exchange of gazes. In this way, the eyes are the messages of the interior: They reveal a You. This cannot be avoided. Only by making my own interior visible can I recognize that of another. Only by offering my interior do I receive that of another. The German-Iranian artist Pantea Lachin called this interconnection "extended vision": Because the other is in a position to see some part of me that I cannot recognize myself, it expands my own field of vision. It means to see with somebody else's perceptions, in the same way as poets see with words. Precisely this is the transformation of loving sight: It enables the seen to become creative. The Russian poet Marina Zwetajewa writes that to love means seeing another person as God intended them and as their parents failed to realize them.[22]

"Esse est percipi"—being is being perceived, as the Irish philosopher and bishop George Berkeley said.[23] When I do not see others whose identities are dependent on me, this amounts to nothing less than robbing them of their identity and their personal autonomy. This is the cause of our childhood traumas. And though *extended vision* might be a great piece of good fortune in our adult years, we nevertheless cannot lay claim to receiving it. We are not lacking our other half, as Plato thought—in the worst case, we are lacking the happy faith in the possibility of our capacity to be when nobody gifts that being to us.

The erotic is the exchange of reciprocal perception through living bodies. It is the foundational moment of any relationship to reality. The world senses me: In my perception, I am sensed by the world. And I thereby experience myself, sensing, perceived as foreign by the foreign-familiar of my own body, constantly regifted to myself. Therein lies the erotic.

The poetic imagination of the body

Our relationships, when they succeed, are constant celebrations of this connection-in-separation. They are examples of the life-instigating dialectic. In this, they are, as Octavio Paz observes, always tragic, even the happiest of them, even those that are afflicted by the fewest compromises: Death will end them. And death, the incommensurability of the other that contradicts self, accompanies them. Our relationships, when they enliven us, involve a deposit of blind trust in an unknown result. They are the overcoming of the paradox of reciprocity—and in this, if they are giving life, they can only be a blatantly obvious manifestation of this same unsolvable paradox. Successful relationships gift to us, in their happiness and in their tragedy, nothing less than what poetry also makes possible for us: In the midst of the everyday of reciprocal touch, they offer us an exemplary comprehension of the interconnectedness of aliveness. Loving is about life, about what it means to be real.

Conducting our relationships ecologically, in this sense, runs according to simple yet paradoxical principles: It means granting to yourself the full extent of your aliveness by allowing others to exist fully in theirs. Only both sides together allow for transformation, make it possible for the space of poetic imagination for one's own life to open up. The successful relationship is the process of mutual transformation that makes both parties more enlivened.

In the erotic encounter, the fact that we enter into a relationship with one another as we are—exposed, vulnerable, with all of our weaknesses—becomes itself the leitmotif. And this is why the erotic encounter makes clear what it means to be alive. Octavio Paz writes: "Love does not defeat death; it is a wager against time and its accidents. Through love we catch a glimpse, in this life, of the other life. Not of eternal life, but [. . .] of pure vitality. [. . .] The time of love is neither great not small; it is the perception of all times, of all lives, in a single instant. It does not free us from death but makes us see it face to face. [. . .] It is not the return to the waters of origin but the attainment of a state that reconciles us to our having been driven out of paradise."[24]

The erotic is reality in highest concentration, the intensive experience of being real as a body-in-meaning. This is precisely why Octavio Paz can connect the erotic and the poetic so closely; both of them are modes of experience in which the world's genuinely relational quality becomes clear, because transformation occurs within them and produces new things, because identities are shattered and then sprout within them, because here the imagination explodes and that ecstasy takes hold that is otherwise experienced only in spiritual trances.

All of the erotic is the transitional movement from body to imagination. The erotic in a literal sense—bodily tenderness and the ecstasy of the skin—is nothing less than precisely this experience: The experience of being felt and feeling both inside and out. The erotic experience allows one to be near to a body, an outside, and thereby to enter into the most intimate connection with an inside. A tender touch makes the boundary of my body sensible and thereby allows yours to become experienceable, but because this experience is mutual, those boundaries are, while being realized, simultaneously torn down, and the land opens up, such that touching bodies allow one to touch a space beyond the body. Bodily touch in love gifts precisely this: the knowledge that only when I accept that I am merely and completely a naked body, only then am I echoed back onto my own physical space, only then does the world become more than body—not just I, but especially You.

The erotic is thereby an experience of transcendence. But it is not a transcendence only of the spiritual I but a transcendence in the flesh. It does not allow for a flight from the world but penetrates deeper into it. Insofar as I expose myself to the other in my unprotected, naked aliveness, the erotic provides the opportunity to experience aliveness as a force that uses the body as a vehicle but is not bound to it, even though it cannot get by without it. Erotic is the poetry of the body, as Octavio Paz said: the way in which we act in the greatest possible aliveness and simultaneously grasp what this aliveness means. If love is the practice of aliveness through which I make myself and the other both more alive and more real, then the erotic is the rapture of aliveness in which I grasp that its space is not only that of material reality but a place of transformation and imagination. The natural world is one variety of poetic space, one that precedes and transcends it. Love is another. The erotic is the perceptible transition point, the moment at which the body is completely soul.

So surrendering to the erotic experience also means understanding that the other is not a means to salvation from the misery of an imperfect, ever-painful existence, but the opposite. Love opens the window to accepting reality with all of the resignation that this entails. The couples researcher Schnarch calls this acceptance "differentiation." It calls upon us to "balanc[e] two basic life forces: the drive for individuality and the drive for togetherness."[25]

The French philosopher Georges Bataille compares the nakedness of the erotic encounter with a symbolic death. For him, the "mise à nu," the act of disrobing, is always also the "mise à mort," the act of disembodiment or death.[26] The body, in its vigorous buoyancy and mobility, always conceals the ponderousness of the earthly, a whiff of imperfection and decrepitude. Showing yourself in your fragile corporeality means accepting; accepting yourself; accepting the other; showing yourself in all of your imperfections; confronting shame; having the courage to laugh at this shame; gifting to the other the possibility of seeing you completely, touching you completely; gifting to yourself, through this gift to the other, your own visibility.

In the caress of the naked, totally unprotected body, in its excitation, lies the potential of the chiasm of perception of which Merleau-Ponty spoke. I touch you and must thereby be touchable. I excite you and let myself be engulfed by your excitation, which excites you further: The reciprocity of a heightened experience cannot occur in a more intimate fashion. And yet this intimacy is based on an insuperable chasm between you and me. The erotic is the experience of being brought forth and seen only through the other, and also sensing yourself with every fiber of your being.

At the deepest core of the erotic encounter lies the act of showing ourselves, naked and thereby vulnerable. This vulnerability is our dependence on the other. It is the thing that we cannot master alone. In this, it is the sphere of our death, our helplessness as a vulnerable mortal body to which the other assumes the role of bestowing life. In the erotic union, we expose ourselves to the other as the "Homo sacer," as the Italian philosopher Giorgio Agamben termed the human being in its state of extreme defenselessness. But not that we might be rescued, rather that we each might experience ourselves as the incomparable and unique and uniquely vulnerable individual that we always were, the one that is self and remains always dependent on the other to become self—that we might become the being that this individual is.

Our defenselessness is the excitation, this voluntary exposure. Not because some sort of desire for victimhood is operative here, but because it is an invitation to the other's magnanimity to endow me with perceptibility and thereby bestow life upon me. This nakedness is often misunderstood. The surrender that it entails is in no way passive. It is a gift that seductively elicits a response. And the response likewise entails gifting life, honoring vulnerability through the gift of one's own nakedness.

According to the inescapable consistency of life, this act of existential symbolism sometimes leads then to an actual act of conception, and later, a real birth, the epitome of giving life to self-as-other.

— *chapter six* —

A PLAY OF FREEDOM

*Don't do anything that isn't play. Because it will be
play if you are meeting your own needs.*

JOSEPH CAMPBELL[1]

Recently, in the wintertime, I was seized by high spirits. The snow lay thick; it was one of those cold spells in Berlin when everything freezes under a hard and smooth layer of white. My daughter, Emma, and I were taking a walk. We led our poodle into the little park behind the bend. There, the meadow slopes gently down to a small pond. The city cleaning department uses it as a spillway for overburdened drainage lines, but in the winter, when a frozen white sheet covers the water, this doesn't matter.

The black poodle romped through the white. My daughter insisted that she and I also "have fun." I had to make a bit of an effort, because I felt a little weariness in my muscles and my aging bones (it was getting harder to conceal it). I would have been very happy just to stand still. But in the end I don't regret letting myself get carried away. How often have I simply "had fun" with my daughter during her life and not used some sort of adult concerns as an excuse to beg out of it? Only a few times, certainly, and all of them are among my most treasured memories.

"Having fun" was the code word for rolling together down a hillside without paying any attention to what the ground might have in store for us. Emma had contrived the activity a few years earlier in the sand dunes of the North Sea island of Spiekeroog during a short vacation there, in the sand dunes behind the beach where children actually aren't supposed to "have fun," in the interest of protecting the coastline.

On that winter's day, we let ourselves roll through the snow down to the pond, pulled along by gravity, our collars full of melting ice crystals, our skin wet and red, breathlessly gasping with intensity. Then we piled up a mountain of snow as a ramp for our Olympic toboggan run. Emma had discovered a torn black garbage bag in the shrubbery. A bit of dog urine had yellowed the snow, but we laughed about it and paid it no mind—as I said, this was "having fun" in the city winter of Berlin. And then we sped down the hillside together on the plastic, Emma in front, I behind, the poodle growling behind us in a snow cloud of exaltation. We spun around, rolled through the ice crystals, and then dove with reddened faces deep into the frozen white.

Again, again! My daughter would not be dissuaded. All the spinning made me feel a little nauseous, but the rapture of the swaying and the speed let me forget the dizziness and the cold. We only stopped once we were completely sweaty inside our clothes and completely soaked through on the outside. For the duration of those fleeting moments, everything was in harmony; for a flash, everything was in accord, nothing was missing. We were completely there. We were a ball made of three life-forms that transformed the snow in that humble park into an arena of exaltation. Nobody needed to tell us what to do, or how. That came from the wisdom of our muscles, of our sensory cells, from the wisdom of the crystalline world that held us. We did something completely useless—namely, the very thing for which all beings are created.

We played.

We were completely alive.

Living joy is life as play. Loving the world means playing with it, in it, with ourselves and each other. Such that nothing is necessary, but everything plays a role.

In play, we comprehend aliveness

Life-forms play. Anecdotes report that not only human children become immersed in "having fun" or creating whole fictional worlds. We also know of playfully tussling rat babies, wound-up great cat parents, and elephant grandpas who let themselves be carried away with uninhibited frivolity. According to some biologists, even ants and termites play with one another.[2]

The finding that play is so widely distributed among other beings is harder for evolutionary biologists to explain than almost anything else. According to Darwinism's economical view of reality, wherein everything that exists must be useful, it is hard to explain why young tigers risk their necks out of playful curiosity, or why, for example, a full-grown stag would expend energy on something that did not contribute to the dissemination of his genes. Since even older animals play—and adult human beings, like Emma's father—the long-held "standard" explanation for such frivolous high spirits, which claims that play is a preparatory exercise for activity that will be useful later, no longer has any traction.

Play does, however, make one thing possible: creating relationships. It is not actually training for the serious business of acquiring nourishment or forcefully defending territory, but for imaginative participation in the creative universe and for one's role within it. So play comes very close to being a comprehension of one's own aliveness. Play is an instance of expression and thereby one of the most important manifestations of erotic ecology. It is the model for an existence that is not subjected to functionalism but boldly carries forth the natural history of freedom and expresses individuality.

Cats do not swipe at balls of yarn in order to practice catching mice, but because it is in their nature as cats to experience their own bodily vigor in the activity of hunting. Just as predators take joy in the hunt (dogs stalking after mice wag their tails enthusiastically), every being takes joy in the activity wherein it experiences and expresses its own nature. Young human beings do not just play with puppets (or games of "mother and child" with one another) because they want to study the useful laws of parenthood, but because it is their nature to enter into nurturing and caring relationships. In play they can sound out and alter their own capacities to relate in free creativity.

We can therefore say that play is the realization of being alive, is sculptural work with the raw materials of that "pure aliveness" that

Octavio Paz identifies as the deepest core of our experience. In play children—and whelps of all sorts—poetize their world. The interesting thing about this is that they already have everything they need for this activity. They follow a creative program that they experience as an instinctual desire for play—one that unerringly shows them the right way. Children do not practice their humanity by playing but express it, and thereby experience it for the first time, by establishing an identity. This construction of identity is likewise an exchange that occurs in relationships, through and through. So playing means forging life-inducing relationships—and also life-harming ones, which one also learns to distance oneself from by playing.

In play the paradox of the relationship between the well-known self and the unknown other is restaged. Playing children seek out themes, places, methods with which they can try out well-known things on the one hand and have to take risks on unknown things on the other. Every day, they venture a little farther out on their quests in the meadows, climbing first onto just the lower branches of the tree, then a little higher the next day. They constantly seek out edges and zones of transformation—like the ones that form where the woods become meadow in the border region of the blooming hedge.

Children are themselves the essence of the living. They are prompted to be aimlessly creative in that undetermined zone between risk and security. In play they constantly define life as the creative transition between control and uncontrollableness. The fact that this is life can be directly experienced and becomes visible in the intensive absorption of their concentrated expressions, which can turn into rolling laughter in the blink of an eye.

Play thus reveals itself as a practice of loving the world.

Love as a practice of aliveness

Astonishingly enough: Our children are masters of this practice from birth. This insight contradicts a tradition of thought that is still influential to this day: that children are unbridled and wild and must therefore be groomed—raised—with force and severity. The grandfather of depth psychology, Sigmund Freud, assembled a terrible list of immature psychological stages that we go through in childhood (for example, the *anal stage* followed by *polymorphous perversity* followed by the *oedipal stage*). Freud was convinced that a human being, at the beginning of his or her life, is as fundamentally gruesome and dangerous as the rest of the natural world. He therefore thought that everybody had to learn to "sublimate" their drives for sex and destruction into culture. With discipline and order, if necessary.

With this, Freud founded a particularly influential variant of the ancient Western understanding that deeply mistrusted life and its forces that sought health and fulfillment. Subliminally, we still conceive of children and their needs, in many ways, as part of a threatening wilderness: impenetrable, hard to control, and necessary to tame, no matter the cost. Immediately upon being born, young people are subjected to our drive to get away from all contradictions. This process also follows our civilization's unconscious conviction that death can be mastered by technology. Children mutate nowadays into success-oriented projects. Our child-rearing is supposed to arm them as best as possible against any misfortune. Children should, so to speak, achieve symbolic immortality by gaining proficiencies, assertiveness, and motivation.

In our fervor, we overlook the fact that children generally know better than we do what it is to be alive. We likewise overlook the fact that nothing can truly rescue us once we have unlearned what it is to be alive. When we forget to orient ourselves toward those foundational principles according to which creative reality constantly unfolds, all of our projects are endangered. Yet aliveness is seldom a goal of child-rearing. Child-rearing is, on the contrary, "give and demand": we constantly want something from our children so that they become better. But we cannot recognize that they have something to give us precisely due to the fact that they are already perfect as living beings. The child's gift to us is the knowledge that people already have everything they are looking for. We simply must not allow it to be taken out of our hands.

That precious thing that we already have at the beginning of life is our love for the world. In play the child loves the world by not only reconstructing it, fascinated and in awe, but by adding new creations of its own.

And children love themselves in the world by savoring their delight in it and allowing themselves to be inspired. A child is born with everything that it needs to participate in the world's interrelatedness by means of self-generated creative relationships. A cosmic genius thus inheres in the child's capacity for play: The playing child places itself in the position of a universal force that constantly creates new connections, varies existing ones, and thereby helps the whole express and experience itself more deeply. All of us do nothing less than this when we play—the human whelps, the ants, all of the interwoven species of our vast ecosystem.

Children come into the world and know how to produce their identities through acts of interrelation. Isn't that wonderful? Shouldn't we, in our civilized isolation, take this up as a positive example? Unfortunately, we mostly do the opposite. We continue to believe that we have to teach our children the crucial things instead of accepting that they already know them, whereas we adults, on the other hand, have unlearned them. So it is a matter of allowing children to rejoice in this knowledge and claim their wholly individual way of being in relationship with the world. Children only need to be given a few elements of our cultural code in order to do this more fully: written language, mathematical conventions, technical capabilities.

Children are capable of playing. And children are capable of loving. And this is precisely what many adults most deeply abuse in their wish to educate and raise their children: They prey upon their children's aliveness in order to strengthen their own, which they lost because their own prior connection with the creative world was snatched away from them. Many parents do not love their children but let themselves be loved by them, with all of the desperation that a child can summon in his or her effort to help a person who actually was supposed to protect this very same child.

Almost nobody has delved deeper into these relationships than the Swiss childhood researcher Alice Miller. She arrived at the insight that the trauma people suffer is the loss of their aliveness in childhood. The secure knowledge and intuitive confidence that their own impulses accorded with the creative world and were deeply correct were driven out of them. As a result, these adults are no longer in a position to act appropriately according to these impulses. Aliveness means being in a position to play with one's own possibilities because one is secure in the knowledge of being sufficient, because one feels allowed to assert oneself as an individual part of the whole creative cosmos. With precisely this, with this assurance, a small child can hold the adults around it transfixed. And this assurance can also be taken away from a child so easily.

Playing means enacting this assurance to be a creative part contributing to making the whole flourish. It is a highly imaginative and extremely serious way of grasping the "terms of life" that Schnarch speaks about, a way of exploring them by transforming them into one's particular, personal style. Playing is the child's method of embodying its own role in the ecology of relationships. The role of the child's caregiver in this process is likewise to follow the terms of life. To do so, he or she must enter as a partner into a relationship that allows the other to become more real, meaning it allows the other to become what he or she already is and therefore yearns to be. The caregiver must help the child transform itself constantly—not into a paragon of parental wishes or fears but into what it already is, even if this still slumbers, enfolded, within it.

"Mental death"

The failure of our civilization in the ecology of loving relationships overlaps with many parents' failure to give their children the necessary space for this transformation. They fail because their own fear of aliveness, learned in their childhoods, sucks greedily at the young aliveness of their children, preventing them from establishing a sympathetic connection to the ecosystem of reality.

As a result, something crucial fails to occur. The reciprocity through which two partners gift more life to one another is consistently destroyed. The stronger partners—the formerly tortured parents—exploit the weaker ones—the trusting children. They sap energy from the life force and the trust in life with which children come into the world. They get rich on their children's willingness to give unconditionally. Because the children cannot recognize that their caregivers are responsible for the lack of generosity, they unconsciously blame themselves and begin to regard the stirrings of their own aliveness as bad and harmful. Gradually, they believe that they are not actually alive anymore, that only the others are alive, that they are dead on the inside. As they age, they increasingly internalize the modern myth of a world that is deeply dead, because the agonizing awareness of their own nonexistence can best be concealed and hidden within that world's requirements and distractions.

Miller observes: "Every child has a legitimate need to be noticed, understood, taken seriously, and respected by his mother. In the first weeks and months of life he needs to have the mother at his disposal, must be able to avail himself of her and be mirrored by her."[3] Miller refers to the British psychologist and researcher of play, Donald Winnicott, who observes in the relationship of mother and child the kind of reciprocal creativity that I analyzed in the last chapter: "The mother gazes at the baby in her arms, and the baby gazes at his mother's face and finds himself therein," Winnicot says, "provided that the mother is really looking at the unique, small, helpless being and not projecting her own expectations, fears, and plans for the child. In that case, the child would find not himself in his mother's face, but rather the mother's own projections. This child would remain without a mirror, and for the rest of his life would be seeking this mirror in vain."[4]

Only this sort of trusting being-in-relationship allows children to really feel their physical and emotional needs. If they experience these needs as unwelcome, then they go the way of a trauma victim, a hostage, an abductee—in other words, those who experience that their wishes for escape and for freedom lead to life-threatening situations. People in such circumstances learn to eliminate these feelings from their consciousness. To mitigate the fear of death, it helps to recognize the human side of one's kidnapper and ignore the violent side. At some point, prisoners can no longer grasp the full complexity of their guards and perceive them as either good or evil, depending on how they act—a kind of black-and-white thinking that makes it possible to survive but not to be alive. But shutting off certain feelings leads to the alternative of experiencing oneself either as simply functioning or rather as inadequate. One's own genuine identity falls by the wayside.

The Australian trauma researchers Angela Ebert and Murray Dyck refer to this experience as "mental death."[5] This is all the more true for little children, whose fragile identities do not have the means at their disposal to receive assurances from the world when they are denied them by their closest contacts. Infants do not read poetry or self-help books. Ebert and Dyck describe the collapse of personal autonomy and the vague feeling of being damaged as symptoms of "mental death." Mental death means the death of feelings—and with that, the abdication of one's own aliveness for the sake of naked survival.

Being alive—creatively expanding the network of relationships through reciprocity—means being allowed to feel everything and thereby not being forced to weed anything out. Aliveness means being allowed to be completely real, in all dimensions of existence. But this is precisely what is hindered by the traumas of the childhood years. They destroy aliveness—and are thereby the echo of our world, which above all else lacks a deep understanding of aliveness, and the courage to recognize it.

"Toxic" relationships

The trauma that many experience early in childhood is a catastrophe of broken reciprocity. It is ecological because it impairs our capacity for connection, which is the basis for all other capabilities. And it is existential because it endangers the continued life of those affected by it. The tragedy begins when we pay no attention to our own aliveness. When we instead listen to what others say, or to the echo that others' decrees (those of our parents or our teachers) leave behind in us. For the individual, this results in a feeling of worthlessness and also a constant effort to conceal this imagined worthlessness. We obediently deny our own legitimate needs and thereby switch off all the dangerous feelings connected with these needs. As a result of these traumas, those affected by them substitute in a better-looking construct—and they also do everything they can to prevent those substitutions from being uncovered.

Susan Forward uses the term "toxic" when referring to the effect of people who do not act on behalf of aliveness but seek to whitewash their own fear of it[6]—toxic like a mercury solution that poisons a river and kills its many inhabitants. When people are toxic, it is always a conspiracy against aliveness, including their own. Somebody acts toxically when he or she sees me not as I am, or as I attempt to be, but instead sees me filtered through his or her own fears and expectations. People are toxic when they consider themselves deeply inferior and therefore wish to control everything simply so that alleged truths about themselves do not come to light. People are toxic when they are not only afraid but also allow themselves to be led by the nose by these fears, and then unconsciously blame others in order to cover up their panic.

Psychologists attribute such behavior to personality disorders. These disorders, along with other torments of aliveness, like depression, are consistently on the rise. According to statistics from the WHO, one-fourth of all the residents of Earth (and counting) battle with emotional-mental problems at one point in their lives.[7] Our relationships are at the center of these ailments. We can therefore say that nowadays the great pains of love constitute a widespread epidemic.

Personality disorders are illnesses of the emotions and thereby pathologies of aliveness and of living feelings. Those afflicted by them cannot sustain the contradictory richness of a relationship but must forcefully produce definiteness and black-and-white contours: control at the cost of truth. Some researchers claim that up to 15 percent of the Western population is afflicted by such disorders.[8] All who suffer such disturbances are united by the trauma of an early destruction of their emotional capacity for relationships: They will do almost anything to conceal from others their feelings of inferiority. They deny their problems so effectively that they can only perceive their own destructive characteristics by ascribing them to the other. Someone else is always to blame. And it is easiest to assign this blame to one's children, because they will not rebel against it. They will do anything to iron out that guilt, to prove themselves worthy, to be seen, until they are finally so broken that, as adults, they will rob their own children of aliveness.

The childhood researcher Christine Ann Lawson has observed that an emotional disturbance in childhood that plants the seed for the destruction of a person's personal, self-affirming aliveness will usually be passed along to that person's descendants. Those whose aliveness has been trampled will desperately use others in order to enliven themselves. The poisoning of aliveness among human beings in a family and in schools is a quiet tragedy. Many assume that it cannot be any other way. They shrug their shoulders so as not to be too intensely reminded of wounds from long ago, suffered in silence, that have still not healed completely below the scar. Concerning their own aliveness, they conduct themselves in a manner similar to how they act toward others: They try to make themselves believe that the trauma is not bad enough to require real examination, and they seek distraction.

Emotional capitalism

The individual dilemma is also so difficult to see because the official way of interpreting reality claims that our feeling is a chimera. The currently dominant picture of life accepts the purportedly universal drive to function better than others in order to survive in the war of all against all. Our society has made the continuous violation of aliveness into one of its defining principles. The avalanching destruction of other forms of life reflects how severely individual emotional life has been destroyed. What makes the current consternation of reality so virulent, and thereby into a candidate for the emotional mainstream, is the fact that our whole climate of living and thinking enshrines a narcissistic worldview of exploitation. The obliteration of feelings does not have a solely private pathological dimension. Personal narcissism is also an echo of cultural narcissism, and vice versa.

Our individual psychological doom manifests collectively as a universal ecological calamity. Both are interwoven with one another. Both are symptoms of a fundamental misunderstanding of ourselves and of reality. In a world of ruined aliveness, it is impossible to be alive oneself. Both inner aliveness and the aliveness of the natural world are facets of something deeply interconnected. As such, the I is indebted to the You, the health of our loving relationships guaranteed only by the ecological health of a world that is constituted by those same relationships.

The blindness that prevents us from seeing this connection can be referred to as "emotional capitalism."[9] Emotional capitalism entails both emotional and ecological abuse. Emotional capitalism transforms the world into something dead in order to evade one's own death. It is a misuse of reality as a disposal product, void of all relationships. It leads directly into the prosperous hell of substitute satisfactions, suggesting that one can erect a small, comforting shield against death.

Emotional capitalism means refusing to accept your own mortality, means denying it at any cost and thereby accepting, or even taking advantage of, another's death. The Parisian psychologist Marie-France Hirigoyen calls such behavior "perverse." "Perverse" comes from the Latin verb *pervertere.* The word means "to overturn" but also "to obliterate." Perversity is behavior that is turned against the path of aliveness. Perversity is the refusal to understand and to accept the dilemma of existence—namely, the fact that we exist in a middle ground between vulnerability and creative power and that our aliveness always includes something of both. Perversity is not recognizing your own weaknesses but looking for them and battling them in others.[10]

Emotional capitalism leads to an isolation of the individual caused by traumatic errors. It absolutizes this experience into a worldview in which separation, the fight for survival, and violence are fundamental constants. It reiterates a mantra of total disconnection that allows everything and everyone to be harmlessly turned into a resource. Emotional capitalism asserts that misuse is unavoidable and compensates for it by promising that the salvation from all evil will be achieved in the near future. It occurs in gestures of possession and of avarice for more. It cultivates feelings of being in the right, not practices of listening. It seeks, in the words of psychologist Marshall Rosenberg, to make violence enjoyable. Emotional capitalism means making the impossibility of love into a guiding principle of life and also preaching this impossibility as a scientific description of the world.

In a world in which a loss of differentiation is the order of the day, in which the natural creation around us is constantly impoverished, "mental death" becomes a diagnosis that threatens all of us. Like some people who are not able to offer their partners a reciprocal relationship, thereby poisoning them invisibly, we, too, let ourselves starve inwardly by clipping the ties to life all around us. Of course, when we have no relationship to all of the countless species that nourish us, satisfy our feelings, and help us understand our corporeality, we lock ourselves up in a traumatic prison. But like all true trauma victims, we have no real idea what is causing our suffering. We try ever more to control it, even though this is precisely what makes everything worse.

It is difficult to answer the question, how a society made up of many individuals whose utmost desire is to control their feelings of self-worth can be exhorted to take account of the principles of aliveness. It demands a great deal of nerve and also the willingness to accept the risk of personal failure. Trying to be real here means at the same time acting politically: Having the courage to feel your own needs and also to trust those feelings amounts to having the guts to stand up publicly and demand a different politics, one suitable to the demands of aliveness.

Both are connected with a great inner necessity for those who have been traumatized and are familiar with the terms of imprisonment. Standing up in public like that requires you to give yourself the right to be. And precisely this—granting yourself the right to your own individuality without any permission "from above"—is so difficult when your own life did not begin with the reciprocity of a living relationship.

Those who would advocate for the rights of aliveness in the world must first award themselves the right to be alive. They must practice looking at themselves with love and accepting their own needs. Only then will their efforts succeed in not passing along a trauma, in not denying to others the right to their own experiences for the sake of the "proper cause," and in not being willing to sacrifice other things and other people for the sake of the "good." Those who would fight for aliveness must first fight for their own feelings.

The barometer of our emotional ecology

Our feelings are the dimension in which we are "inwardly" shown whether our ecology of mutual exchange is fruitful. Consequently, they have a quasi objective character and are never just personal fantasies. Feelings are the barometer of aliveness within us. They are the inner form of the outer state of a stretch of landscape that we can directly comprehend through our senses: a blooming meadow, a clear-cut wood. Feelings are the meaning of our inner life circumstances. They are what we are—not what others are. Susan Forward encourages us to trust our view of this inner landscape. It is also always right, just as trees and flowers cannot lie. "No matter how confused, self-doubting, or ambivalent we are about what's happening in our interactions with other people, we can never entirely silence the inner voice that always tells us the truth," writes Forward.[11]

Feelings are the voice of truth about our existential position. They are subjective because they deal with our flourishing. But they are also objective because they express the measure of our aliveness. They are the expressed experience of our own ecological equilibrium. Or to put it more clearly: Emotions are a poetic commentary on our own existence, just as indirect, just as creative, just as difficult to suppress as an outcry, a drawing, a verse, a melody, a landscape, whose emotional substance shakes us to the core.

Feeling is the partisan squad of the individual I, a group that is always in the know and must therefore be muzzled by all oppressors. "These people have all developed the art of not experiencing feelings, for a child can experience her feelings only when there is somebody there who accepts her fully, understands her, and supports her," says Alice Miller.[12] But without feelings, in which a person's own needs can be expressed, those needs are without a voice. Yet only when we listen to and know and recognize our needs are we truly connected with our true, living selves. Only then can we trust our own perceptions, for they show us what truly is.

On the other hand, people will form a "false self" as adults if they were made to feel, as children, that the healthy experience of their own aliveness and their natural needs threatened them with annihilation.[13] They subconsciously strive for the identity that they were allowed to adopt, rather than the one that slumbers within them. This is also a creative process, of course. It unfolds with the same resilience with which the natural world transforms catastrophes into stories of survival. A river that has been flooded with toxic fluids is still a river. It becomes a "dead zone"—it lacks almost all oxygen. But still it is not truly lifeless. It transforms into a body of water with low diversity, where only algae, bacteria, and rotifers can flourish. The river preserves its connection to aliveness. To be sure, it exists at a much less differentiated level than it did before the poisoning. Viktor von Weizsäcker recognized that even in a moment of brokenness, the forces of creativity prevail; as he said: "Every sickness is an incomplete act of creation."[14]

These creative acts are always expressed as bodily gestures. The natural world conveys to us, through such gestures, its inner conditions. Unlike a human being, the natural world discloses its existential constitution to us without shyness, shame, or ulterior motives. This is not a conscious process, but at the same time, such experiences are not concealed. Monotone plantations and colorless acres speak volumes about the extent to which creation is no longer active within them. Nature can thus be defined as a space in which the experience of aliveness is revealed. In it, the actions of plants and animals always show their feelings honestly and forthrightly. They are evident in their bodies. They continually allow us to recognize how they are getting by in life, whether they are flourishing or doing poorly. They appear dried up, are shaggy, full of flowers; they carry themselves feebly or appear in powerful flight. As such, they are acts of creation that have become visible, that stand before our senses, that invite us to partake in them so that we might feel our own feelings again.

In the web of life-forms, everything that we need to know for the unfolding of a healthy identity searches for its expression. This is one of the significant reasons nature is able to offer us comfort and support, whereas we are so keen to hide these experiences and the feelings that accompany them from ourselves and from others. Particularly for children and adolescents, such a frame of reference is existentially significant. Nature is fully alive. That means that it shows all its truth without shame. Seen in this way, the natural world is the ultimate reality that can embolden us to take the same risk of truly being. It is the place that offers comfort when our needs run up against denial or ignorance. It is the space that conceals no feelings and therefore allows us to admit our own feelings. In the inside of nature there is no unsuitable or false aliveness. The natural world is further removed from totalitarian action than any form of human achievement could ever be.

Nature grants to everything its authentic self. Our ecological relationship thus becomes the prerequisite for the principles of personal emotional authenticity. These principles are precisely what children attempt to unfold in reciprocal exchange with their caregivers, who often forbid this unfolding. No self-determined life is possible without this space of free emotion. The American psychologist and pioneer of family therapy Virginia Satir calls these principles the "five freedoms." They are imperative in order to claim a feeling identity and to acknowledge the same in others.

The "five freedoms" are:

1. The freedom to see and hear what is here, instead of what should be, or will be.
2. The freedom to say what you feel and think, instead of what you should.
3. The freedom to feel what you feel, instead of what you ought.
4. The freedom to ask for what you want, instead of always waiting for permission.
5. The freedom to take risks on your own behalf, instead of choosing to be only "secure and not rocking the boat."[15]

According to these precepts, a good relationship is always also an ecologically stable relationship: one in which truth is active. "To live in truth"—for the ecophilosopher David Abram this is the actual reason that we might seek to protect a landscape and to make our peace with it.[16] Both sides participate in such a relationship. They can transform themselves within it and do not need to conceal reality. In other words, this also means: For as long as I perceive even one catkin in the springtime, I am working against oppression.

The transformations of love are only imaginable when nothing is withheld. We can only gain access to such transformations, and thereby to the maturity of adulthood, by allowing ourselves to see, to feel, and to express what we experience. If there is no openness for this transformation, love has it tough, because it is the transformative element *par excellence*. Accordingly, the psychologist Abraham Maslow observes: "We have discovered that fear of knowledge of oneself is very often isomorphic with, and parallel with, fear of the outside world."[17]

I remember one early spring day when I had captured two insects in my kitchen in Liguria that had emerged from the stack of firewood. The two black-and-yellow-striped longhorn beetles at first climbed restlessly up and down the walls of the empty honey jar in which I had barricaded them. At some point, I added a bit of clover to the jar. Immediately the two beetles began to mate—they had forgotten they were imprisoned. The blossoming plant had transformed the threat into the ecstasy of the new year.

For the duration of our lives, we struggle to gain the "five freedoms," whereas nonhuman creatures and people in the first months of their lives guilelessly embody them. Their confident aliveness is the ludic model that we can think back on for the rest of our existence. In order to allow ourselves to become ourselves, we have to feel as they do—and as we did at the very beginning—once again. We have to admit our needs and articulate them. When we listen to our needs, we can also perceive those of others. In that moment, the reciprocity foundational to aliveness for all is restored.

Trusting your own death

The natural world is the mistress of a reality of personal experience. But only because she is not a kindly "mother" who reconciles us with everything and houses us under her protective cloak, only because true death awaits everything and because nothing can conceal the genuine possibility of it. Anyone who has ever been lost in the wilderness can empathize with this feeling. Dying is a possibility! Nature does not hide this fact. On the contrary, it sculpts with it: Every currently living species exists at the expense of hundreds of others that have long been lost. The tremendously salutary realism of the natural world is composed of such expirations. The opposite of emotional capitalism and the "mental death" that results from it is not, therefore, the "good mother Earth," where only symbiosis is operative. Rather, it is love as an ecological practice that accepts the presence of death in its midst.

Dying is essential to awakening from the torpor of insufficient aliveness. Becoming alive means experiencing the death that you have feared since childhood, because it was always there at the dinner table as a constant threat to your own aliveness. It is therefore imperative to grasp, in your own life, that dying is actually the part of life that you have sought for so long. Only after we have gone through the death of our own central personal concern will this concern really become legitimate. Only then will it bestow upon us the identity that we seek.

This same courageous dying is something that children should learn from their parents. Yet because parents often need their children in order to survive emotionally themselves, they refuse them the experience of this good death. I am not referring to actual death, of course. I mean situations in which the terms of life demand that one surrender to them and also convey trust to another with a loving gaze: This situation is survivable. Situations like a fight among schoolmates, a flunked math assignment, an unfinished first treehouse, a skinned knee, the first childhood illness. Supporting someone in these passages through death requires that the caretakers both let go and be constantly present. Instead of protecting, they should help by simply being present with the affected person when something goes wrong.

Good parents allow children to experience failures, because they don't judge and don't control. They enliven their children because they trust. They see their children's living resilience and know that it can get them through things. They enliven because they demand no reward and do not downgrade their children to a means to an end, but offer them existence as ends unto themselves.

Trust, that advance payment on the future, is also something that the natural world gives to us. It is the place that says, "Look, I do not change, I remain the same, whatever you might do with me." It makes no evaluations, but elastically springs back into an accepting state of equanimity, like a young birch sapling in spring that children bend to the Earth in their exuberant play.

The paradox of the "principles of life" cannot be resolved; it can only be personally lived—and personally died. There is no path into the ideal world. The rift is irremediable, because it is a component part of creation —indeed, it is the very thing that made creation possible.

Leonard Cohen got to the heart of this in a few short lines in his song "Anthem":

Forget your perfect offering.
There is a crack in everything,
That's how the light gets in.[18]

The fear of death, on the other hand, never leads us to escape it. Often, it results in our sending others to their deaths. The fear of death beguiles us into refusing transformation. Refusing to accept that death, which is a necessary part of life, brings life to a standstill.

Because we human beings are so afraid of dying, we must constantly relearn a culture of life. But not just as individuals. Nowadays we should recall collectively as a civilization that such a culture is necessary in order to make space for life. Perhaps this is our task during this critical time on the planet. So much dying brings about the forceful realization that due to our fear of death and our lust for eternity, we have repressed things that other societies at other times perhaps understood better than we do. A culture of life entails a constantly renewing creative freedom with respect to the inevitable necessity of death, in order to share the world with one another as feeling bodies bound to the terms of their biological physicality. In this culture, two things are important: acting freely, but also accepting necessity. Only a culture of life has the chance to develop the same acceptance of failure and conducive austerity that is constantly self-produced within ecological reality. Ernest Becker writes: "If neurosis is sin, and not disease, then the only thing which can 'cure' it is a world-view, some kind of affirmative collective ideology in which the person can perform the living drama of his acceptance as a creature."[19]

Merely to think about the possibility of such a culture nowadays is, at the very least, a task not only for individual emotional survival but also for the continuation of our collective existence as a civilization. A culture of life could provide us with that piece of ecological insight that we lack by nature, because we have more freedom in our actions than every other biological species. Other species' aliveness is part of their natures. We have to freely choose to engage with our own aliveness. The necessity of culture thereby shapes our natural condition. And this is precisely why our nature as a culture must include the body and its unfathomable capacities for expression. Unlike our companions in the animal and plant kingdoms, we can only live the natural state by recognizing it and actively making it into the goal of our actions. We recognize our nature by uniquely restaging it in our own state of freedom.

For this, we must freely choose to constantly recreate the principles of life that are indissolubly bound to our bodies. Because culture is the creative staging of our natural state, it cannot be allowed to pretend that we are above the principle of living existence and the terms of creative relationships. The behavior of a species is a variation on the necessary principles of its aliveness in the context of other living beings, a negotiation, a creative implementation of possibilities. Within this limited framework, a culture of life could unfold. It is creation in an interplay between our self-instantiation and the principles of reality, through which both dimensions can be more deeply and clearly experienced.

Considering the situation of the planet, we should do everything possible to explore how such a culture of life could be provided for. These considerations can bring a new motif into play: the acceptance of being real, of just being what there is, and with it the productive determination to renounce all appeals to the superhuman or to that which is outside of reality. The dominance of Western thinking during the last millennia has surrendered culture all too often to utopias of salvation. It has, however, barely sought to generate the compromises that would be required to become fully enlivened within the imperfect conditions of reality. But if culture denies the mortal share of our existence and seeks to overcome our creatureliness, it is unsurprising that it should scorn and destroy the actual natural world, outside and inside ourselves—for nature is both the guarantor of our vulnerable existence as creatures and a lasting reminder of it.

Ernest Becker says that, "culture is . . . a heroic denial of creatureliness."[20] According to Becker, the central misery of all cultures (not just ours) is the denial brought about by our permanent fear of death. He observed that people of all eras have striven for immortality by various means: the worship of omnipresent ancestors in rocks and trees, the eternal life that awaits all true believers, the technological deliverance that will allow the world to be conquered. There has always been a heroic path to immortality. Following this path has always required strict rules that required one to give up the very thing that was supposed to be preserved: one's own aliveness.

Becker belongs to a school of humanist psychologists that includes Erich Fromm, Rollo May and his previously mentioned colleague Abraham Maslow, Virginia Satir, and also Alice Miller. What all of these thinkers have in common is that they accept the embodied reality of our existence and understand health through the ways in which we invent creative solutions to its inescapable dilemmas, and also in the ways that we fail.

Like his fellow combatants, Becker thereby razes the ambitious edifice that Sigmund Freud erected. For Becker, our deep dilemma is not the repression of our sexuality, but the repression of our consciousness of death. Neuroses and psychoses are, according to Becker, various forms that manifest the suffering of people who have lost the courage to advocate for their own reality in spite of their fear of death.[21] In addition, Becker's colleague Wilfred Bion directs our attention to a critical aspect of this same theme: "The fundamental problem is, how soon can human beings reconcile themselves to the fact that the truth matters?"[22]

Character armor and romantic love

To love—which is to say, to live and let live in the deepest sense—is not a deliverance from death. To love is to welcome death and its mysterious, unending power of transformation and creation. The philosopher Hildegard Kurt writes: "Listening [to others or to the Other, A. W.] begins where I die."[23] All of this is silenced in the pleasant myths of the "warrior of the light"[24] that are currently circulating. They push for aesthetic heroism. They, too, preach a form of heroism to which one can gain entry simply through proper behavior. They conceal what Rilke expressed in his requiem: "Who speaks of victory? To endure is all."[25] But not endurance in the sense of survival, in the sense of some discontented perseverance, but as that oft-described state of drifting, all senses open wide. As openness for the creation that is always the Other, never the Ego.

Love that risks one's own death in order to make space for aliveness is not part of our culture. It remains the heroic act of the individual. The fear of death is, however, so great that people often cause particular offense when trying to be alive and showing that it is really possible to exist openly above the abyss of nonbeing. Of feeling all. The fear of death is so great that emotions appear dangerous and are supposed to take up as little space as possible. Anyone who displays them too naturally is suspect. Anyone who believes in them is dismissed with a weary, ironic smile. Children, of course, often cause offense. Children—wild, naive, unadjusted, thoughtlessly aggressive, boundlessly generous. Children, who fall down and stand up over and over again: In a culture that seeks at any cost to repress all forms of failure as omens of mortality, children are dangerous.

But even adults have it hard if they do not submit to the standards that control risk (both of life and of death) or (worse still) do not exercise such control themselves. Paradoxically, they attract hatred and rejection, even though they possess more of what everyone so desperately covets. But the vivacious, generous person who is friendly without any ulterior motives also demonstrates constantly that he or she has something that others do not. If people's guilt and shame originates, as Becker suggests, in their own unused and repressed life, then the lives of those with an overabundance of aliveness will be sought to be destroyed. Otherwise, their energy would be a persistent accusation, one that would make the repressed suffering unbearable.

The psychologist Wilhelm Reich described this societal hatred of life as "the murder of Christ." For Reich what made the Biblical savior exceptional was not that he enunciated a new religious message, but that Christ was truly alive. He couldn't be anything but real. He was real in an ardent, childlike, and instinctive sense. And this, according to Reich, is precisely why he had to die. Reich concludes that the fate of the historical Christ figure resulted inevitably from the hatred called up by his loving aliveness in those who did not dare to risk it themselves. According to Reich, the repression of aliveness in oneself and others was, and is, a repetition of this first murder of Christ, each and every time.

The uncontrolled repetition of this act lies at the basis of our attitude toward the naked chicken of the battery farm; it feeds our indifference toward the marshland that yesterday opened its watery eye to the heavens and today has been drained to become a cornfield.

The reflex to kill a person who attempts to be alive is the same as the mistrust of an animal, of wildness, of our own feelings. It is related to the need to educate and shape a child. A civilization that shuts out death must control life. It tries to annihilate all that is living and that follows its waking emotions, the "five freedoms" of Virginia Satir, "the inner freedom of the animal man which is part of the lawful freedom of the whole creation."[26] The prohibition of these freedoms, according to Reich, flows out of a single destructive impulse: "You must not ever, under punishment of death, know God as Love."[27]

The structural hatred of the living lies concealed below our "character armor." It is controlled by systems of rules that determine who counts as good and high-performing, as useful and worthy of love. Who is cool and hip and belongs to the inner circle of those with a special value. The hatred is well hidden by the persons in power. It is usually concealed as care for what is best and love for the other. The hatred only affects those who deny control and try to live truthfully. Thus, the particular act of malice against life involves more than the unconscious transfer of a person's own pain, out of the fear of imperfection and of death, onto those others over whom you exercise dominance and control. The true act of malice is no longer feeling the pain and declaring that the acts of hatred motivated by that pain are acts of care and good intention. Hell lies hidden in denial.

We shudder at knowing the extent to which control through self-denial is part of daily life in totalitarian systems. Those who transgressed against Stalinist truths might one day find themselves ushered into a black car at four in the morning by silent men and thrown into prison. But the worst part was that the suspected delinquents had to compose repentant confessions to the party in which they renounced their aliveness and the "five freedoms." Only after they had betrayed themselves, after they had begun, out of a fear of death, to question the extent to which they might indeed have been guilty of it all, only then would they be shot.

Those who try to defy our dominant system of efficiency are nowadays threatened by a less heroic elimination. Without certificates of achievement and patronage, they simply cannot participate. They are destitute, powerless, shut out, perpetual children. And to everyone else, they are doubtless "to blame" for their own misery. These anonymous losers are everywhere. We encounter them in the lost ball of feathers that is the shivering sparrow sitting in a shrub in the last bit of village green in winter, we meet them in the small farmers of India who worry about their livelihoods because they refuse to plant patented seeds and can no longer keep pace economically—or who are facing economic ruin because they did plant patented seeds and put themselves hopelessly in debt to do so.

In favor of a pedagogy of permission

Confronting this denial is our most important task, and also our most difficult one, because we must first expose our self-denial. The rest will probably occur automatically, because the waters of life always find their way through open channels. A culture of life begins with speaking of what is and accepting what is. It follows that a culture of life begins, again, with children. For the most important aspect of children is to welcome what is, what claims existence for itself.

Alice Miller frames this as a pedagogy of permission. For her, child-rearing means doing nothing except asserting one's own boundaries and being an example of what it means to *not* allow yourself to be used. This form of child-rearing would be to demand nothing. What children need to unfold themselves "is the respect of their care givers, tolerance for their feelings, awareness of their needs and grievances, and authenticity on the part of their parents, *whose own freedom—and not pedagogical considerations—sets natural limited for children.*"[28]

Children need the "five freedoms," and they need the infinite trust of others that they will savor those freedoms. They need the confidence that they will be able to be alive. They need their parents' joy that they are as they are able to be. They need their parents' wish for them to completely live their own life. In other words, they need nothing less than an ecological culture of love.

PART THREE

WE

Just as the creation of art is a love relationship with the world, the creation of culture and society is a loving mastery of the ambivalence of self and non-self.

PAUL SHEPARD[1]

chapter seven

THE THOUGHT OF THE SOUTHERN MIDDAY

The ecstasy of love can always be controlled by art.
ALBERT CAMUS, *THE REBEL*[1]

I walk through my Italian village on a night whose air is so soft that my footfalls seem to make no sound. Every surface is rounded, moderated by tenderness. It is the end of April. The moon hangs full and white, the mountains look to be pallid blue, a world reverted to the color of older men (blue suits, blue shirts, blue sweater vests). The nightingale sings down by the river, another answers from a little farther off. All the way to the mountains of Carrodano, behind which lies the ocean, the valley is filled with sonority and volume as though it were filled with solid matter.

The bright blue night transforms into voices. A nightingale here, one in the distance, another still farther off. The entire valley is a rolling wave of fluting cascades and crescendos. The Earth lies still and soft under the moon and the blue edifice of the night. I imagine that one could perceive the Earth from space as a carpet of sound, a blue ocean of nightingale voices, one rolling along behind another all the way to the sea's edge. I could fly from nightingale to nightingale and arrive the next night at my garden in Berlin. They are the sonorous vestment of the night-blue Earth.

Without my involvement, the landscape has matured and ripened, has filled with life and presence. Every minute, the blades of grass grow longer. The orchids vanish under the grass. While walking my dog, I come upon a fat toad: the biggest I have ever seen. I have to use both hands to carry her. The dog tracked her down; she was sitting on the hard tar of the little street that climbs to the cemetery between the garbage depot and the forge. I take the toad in my hands; her long back legs dangle, her soft belly full of spawn hanging over the sides of my hands. What water could she have been heading toward? Around here, I have only ever seen the rapidly flowing brook, aside from the puddle outside the co-op market. I can hardly imagine she was heading there.

I carry the amphibian to the garbage truck parking lot. At the back, the ground descends steeply; somebody has recently dumped some rubbish there. The animal's skin is cold and dry against my hand. I imagine the toad, her belly full of eggs, crawling over the trash before plunging headlong into the brook. So I turn around. I almost make it back to the spot where I found her, then turn around again. Maybe on the other side of the street, upriver behind the bridge, there is a spot where the animal could get down to the water? But is she even trying to get to that brook? It isn't actually good spawning water for toads. But if not there, then where else could this animal have grown up?

The dog keeps jumping up toward my hands and sniffing the fat amphibian. For at least fifteen minutes, I walk back and forth. In the end, I give up. I resolve to put the animal back where I found it. The toad will know best where she has to go. An SUV goes barreling past me, its high beams on full blast. The animal in my hands paddles weakly with her feet. At least the car didn't smash this toad into a bloody pulp. Back at the spot where I found her, I discover another toad, perhaps half her size. It is facing down the street. I set my animal beside it. The little toad quickly hops onto the big one, then the two disentangle themselves from one another. The dog sniffs again.

The next day, there is nothing to be found on the street—no bloody trace of an accident. Somehow, both of these creatures, surrounded by stones and tar, apparently made it to the right place. What allows such an amphibian, with such a limited viewpoint (its "frog's-eye view"), to safely find its way, like a sleepwalker, between the asphalt and the high walls? For as long as I had held the toad in my hands, I had been filled with a sense of confidence. My skin nestled against its coarse exterior as though in some tentative exchange of tenderness. And I was filled with gratitude for such trust.

As I returned home on the evening of my pointless rescue attempt, I perceived the tranquility around me. The scent of nameless flowers sweetened the air. I heard the owls call. Their voices seemed to come from every direction. There were three of them—one nearby on a hill across the way, another farther off near Centocroci-Pass Street, a third in the mountains above me behind the cemetery. Silently, the grass stood tall in the darkness, little flowers peeking from among its lanky blades. I suddenly understood that the plants do not stop growing at night, that they do not sleep but strive ever onward, like ships plowing through the pitch-black ocean while all but the bridge crew are slumbering.

All at once, the meadow seemed to me full of individual characters. Every blade of grass, every calyx—a self. To me, the world seemed dispersed into countless sentient individuals, each of them a subject trembling with experience and feeling. And all of them, like me, had only their allotted time, a narrowly delineated space of limited possibilities. There were few options, and one had to reconcile oneself with many things. In the case of the meadow, for example, the grass had to be grazed (or mowed) in order for the meadow to maintain its status as meadow. If no one were to cut the grass, then within a few years the first shrubs would crop up, followed by trees, and the impenetrable Ligurian wood with its hawthorn and cherry thickets and solid oaks would conquer the hillside.

A meadow must be consumed so that it can be itself. But at the same time, it cannot be destroyed by this damage. It grows on it. There is a very narrow region where defamiliarization and estrangement—necessary, yet hardly bearable—overlap. And this narrow zone is the place where all who take part in the meadow's existence—the blades of grass, the crickets, the fireflies, the sage, the orchids, and the lizards—are real. In the velvet of that first early summer night, I walked that slender ridge. And because I was touched by everything in it, it became my reality as well. I understood that all of these beings could only live fully because they all held fast to their singular existences and expired in pain, thereby renewing themselves as part of a collective whole. This alone was what enabled them to also enliven me, joyfully.

It became clear to me that all of this beauty did not result solely from a battle for survival. It exhibited no signs of conquest, but also no signs of surrender. What was happening here had little to do with these superficial categories. This beauty was an act, a practice. It emerged at a level of reality where it made no sense to speak of conquest and surrender, of great achievements or inadequacy. What I saw, what absorbed me, was something like an enormous compromise. The sense of accord shuddered with every gust of wind before the state of equilibrium was restored. The momentary result of this success offered itself up to be enjoyed. It waited expectantly for every fiber of its being to be fully savored—also by me.

Its beauty was a middle way. It was apparent that this way did not arise from the boring and lifeless disavowal of differences. The dark sea of grass blades was vibrating too much in the impatient night air. On the contrary. The meadow's compromise was a gesture of tremendous tension between excess and disappearance. It bound together the two principles of reality—death and totality, collectivity and individual destiny—such that both became very clear and neither dominated the other. The meadow lived the poetic solution to the problem of how to close the open wound of existential contradictions. It held open the full potentiality of both sides and was not open to corruption. Its gentle pulsing under the canopy of night was the wound that was continuously closed, even as it reopened in a different spot.

The meadow is the circumstance of existence that allows both poles—nothingness and plenty—to constantly transform one another.

Neither victim nor hangman:
the art of ecological living

"Men are grass": The meadow is like us, because it is an embodiment of relationship. It is composed of dozens or even hundreds of species, and of what these species create through their interaction with one another. Worms, mites, and other earthly invertebrates are among the participants, as are diverse species of sweet grasses, lungwort, and orchids; primrose, marsh pugs, wild carrots, and cornflowers are part of it, as are ants, beetles, flies, bees, aphids, grasshoppers, blues and coppers, swallowtails, voles, green woodpeckers, and people. They all can only exist in the meadow in a manner befitting their species and behave only according to their inner drives. But at the same time, the bodies of all the beings who compose the meadow have extremely narrow boundaries placed around them by the other participants in the meadow's existence.

The "principles of reality" hold light and dark in balance. Our problem is that we too often seek refuge in one side and deny or kill off the other. If a practice of being in relationship involves the creative transformation of the opposing poles that are necessary for that act of transformation, then both poles must equally continue to exist. Only then can they enhance one another and allow reality to be spoken of with greater intensity.

The question of how one pole is transformed by the other thus becomes a question of moderation. How are we to live as both individuals and in relationship? The essential ingredients do not all taste good alone, but they are vital in certain combinations to the production of enjoyable, excitingly delectable dishes. The proper balance of opposites is found when, in the context of some new breakthrough, their contradiction becomes a creative imagination that is understood not as a solution but as a new and beautiful complication in and of itself. This is why the proper balance cannot actually be measured. It can only be recognized in the sparks that set it alight with enlivenment.

So it is a question of proper balance.

The meadow offers its answer. And on summer nights such as this one, we are completely enraptured by it. But for our own lives, we must find our own answers.

The concept of moderation is not new. It was a key part of ancient ideas, at a time when philosophy was understood less as an analysis of thought and more as an art of living. And this is also our intention here: Erotic ecology strives for thought that conceives of us not as analyzing machines standing apart from the world, but rather as beings that are in ongoing material exchange with the world, constantly framing this exchange in ways that benefit both ourselves and the surrounding relational network.

Science that attends to this exchange automatically becomes a *practice of knowing*. And ecological knowing enlarges into an *ecological art of living.* Erotic ecology would then be an art of living in which existence is enabled by other beings that touch us physically and establish our horizon of meaning. The art of living is the art of being alive, and life is an artistic process. The horizon of this art is, in the words of the French philosopher Michel Onfray, "a life practice that says 'yes' to life (and 'yes' to all of its contradictions) but 'no' to that which destroys it."[2]

The idea of proper balance and the idea of a practice are paired, for the act of negotiation is foundational to acts of moderation. The proper balance cannot be decided in advance. If it were, then it would not be an act of moderation, but an assertion of principles. Of course, such principles dominate the classical teachings of right action—moral laws such as the "categorical imperative" of the Prussian philosopher Immanuel Kant. These are seldom the teachings of an art of living and are rather categories of duty that consider good intentions, but not their outcomes. Acting in moderation means being guided not by principles, but by enlivenment. Enlivenment is also something different from the utilitarian principle of the "greatest good for the greatest number." A practice of the art of living must be in a position to make the tragedy of existence into one of its central components without appointing itself judge and jury over the destructive elements. It cannot manage this without the poetry of bodily touch, in the form of both pleasure and pain.

The most significant modern European thinker who tried to come up with a thinking of moderation was Albert Camus. In his view, the shining light of constant birth offsets, but does not repair, the tragedy of creation. Of course, Camus was not an explicitly ecological thinker. When he died in a car accident in 1960, the environmental crisis was only beginning to slowly seep into human consciousness. In the mind of this French thinker, civilization's most difficult dilemma was that political movements were using the promise of future salvation to justify brutish violence against people. Since then, economics' promises of salvation, which have been paid for in the death of the natural world, have enlarged the matter into a metaphysical catastrophe.

Camus does not take sides in this conflict. Nowadays, we would say that he is neither right nor left, neither libertarian nor green. When faced with this choice, Camus foregrounds a deeper question: How do we manage to be "neither victim nor hangman?" At the beginning of the twenty-first century, this is the central question of political ecology, more important than any other to the continuance of life on this planet.

Camus attempts to answer it in a way that his contemporaries do not understand. He confronts this question by assigning a different status to reality. For Camus, reality is not an imperfect thing that we human beings must change using technology and utopian ideas. It is rather an imperfect thing whose imperfections turn us into collaborators in an act of joyful creation. This is why Camus is a key figure for the future thought of a global culture of life.

For the idea of moderation, Camus invoked the ancient goddess Nemesis, who punished all dogmatic exuberance. Nowadays, Nemesis manifests herself in the parching heat storms over the Great Plains of Australia, in children's blank expressions as they fix their eyes on little touchscreens. She is the goddess who does not condemn innovation but will take vengeance for exclusivity and all forms of dogmatic single-mindedness.

All political philosophy begins for Camus with becoming aware of one's own aliveness. This aliveness is both the everyday integration of inconsistencies and also the opposite of all utopias—in other words, it is the whole of reality. As bodies that bundle together the matter that passes through them with an identity that is at once unchanging and fluid, we life-forms are paragons of moderation: we *are moderation*—we are living forms that exist in an area bounded by beneficial and destructive elements, in a space of moderate energy input and limited energy output.

In Camus's understanding, acting in moderation means recognizing that virtue can never be separated from reality without becoming a principle of evil itself. The desire to avoid all bad things becomes a bad thing itself. But not avoiding bad things implies allowing bad things to happen. It is not possible to solve—or rather, to soothe—this dilemma by thinking abstractly or establishing political principles. The only successful way forward is through an act of bodily imagination—through a transformation, which is to say: through the element of poetry. As expressed in the ecological dimension of the Ligurian meadow, acting in moderation means integrating birth and bereavement into a whole that contains both but creates something greater than either. Establishing the proper balance is thereby an enlivening act—an imaginative act, not a bureaucratic one. Establishing the proper balance is an act of transformation. And transformation is imagination.

The voice of this imagination is certainly not abstract. One can only participate in it. The need to achieve a compromise sounds sterile on paper, but a community such as a family or a village can fill something that is imperfect with life and imagination, such that its enlivenment catches and spreads from the inside out. There is a poetry of participatory sympathy, of co-feeling through sharing and mutual moderation. Indeed, there is no other kind. Poetry, too, depends on moderation—there is no work of art without form, no poem without the rules of language and meter, no music without the fixed relationships of tonality, no visual art without the preexisting relationships of volumes, graphical forms, and wavelengths of light. And at the same time, poetry's gesture of creative, infectious, and inspiring enlivenment, full of life and mystery, renders all measurements obsolete.

Acting in moderation means not replacing the imperfections of creation with worse copies that are ostensibly error-free. By enduring the imperfections of creation, we are already following the goddess Nemesis. The errors intrinsic to all created things can only be endured if we respond to them, and create. As such, the experiences of the embodied present become the strongest argument for the conviction that the world, in all of its bitter imperfection, is the expression of something that cannot be improved. "Perhaps there is a living transcendence," writes Camus, "of which beauty carries the promise which can make this mortal and limited world preferable to and more appealing than any other."[3]

In his reflections, Camus touches on the principles of life: How is an integration of contradictions imaginable without a violent resolution? How can opposites be kept open and brought into creative tension such that one pole does not engulf the other? "The world is not in a condition of pure stability," Camus observes, "nor is it only movement. It is both movement and stability."[4] And thus reality exists in a state of painful tension whose next leap can never be controlled. Its potential can only be transformed by a new act of creation, just as the regular cutting of the meadow transforms into a blossoming of diverse species that twist and twine about one another. Citing the thinker Friedrich Nietzsche, Camus calls for the following, as the central premise of moderation: "Instead of the judge and the oppressor, the creator."[5]

Balance that holds both poles in abeyance can only be imagined as the transformation of one pole through the other. And such transformation is a genuinely creative act—this, too, like the gesture of the nocturnal plant stalks in the Ligurian mountains, is a poetic solution. It is poetic not because it adopts an aesthetic posture, the position of the flaneur who observes and comments without ever getting involved. No, it is poetic because it genuinely demands a creative act and because every creative act creates something new, rather than simply preserving the status quo. In flashes of creativity, these acts can transform misery without abolishing it. Just as cells transform the constant deprivation they suffer—due to the lack of matter and the quest for nourishment—into the complex experience of a world.

Love in its current form—love as a consumable and yearned-for destination or a hedonistic anesthetic—is the strongest antagonist to this idea of moderation, and also the strongest antagonist to poetry and creativity. As it is most often practiced nowadays, love is altogether immoderate, as shown by our immense dissatisfaction when a "relationship" does not offer us everything, when a person does not save us, when a new body begins to show the stubble of personal idiosyncrasies after the first night. Ironically, love is the one area where Camus himself was immoderate, ecstatic in the intensity of his relationships (he always maintained several at once), and unhappy. Camus himself failed to recognize that love is not an act of devouring or being devoured, but is a practice of moderation.

In theory, Camus recognized, as did Ernest Becker, that love is an attempt to undo an absurd fate in which success is always counterbalanced by death. He expresses this in his epochal study, *The Rebel* (*L'Homme révolté*). The book was published in 1951 and contained a poetics of the body full of existential paradoxes that was too ahead of its time for France and Europe. Camus tracks how our dissatisfaction with real relationships has led to the production of systems for salvation, including an impulse to turn the interpersonal relationship itself into an ideal salvation. Camus sees this deification of love as a form of insurgence against reality. Like every act of human resistance against the foundational principles of reality, such an apotheosis of love will fail and ultimately turn into the very thing that it sought to overcome. The only way out is to create, to imagine, and to pass from the revolutionary's urge to destroy to the rebel's moderation. A rebel says neither "no" nor "yes, but." He always answers with a "yes and." He adds instead of cutting away. He enlarges, as Camus observes: "In every rebellion is to be found the metaphysical demand for unity, the impossibility of capturing it, and the construction of a substitute universe."[6]

By and large, people experience the failure of love—and the ensuing urge to destroy—privately, but we are also experiencing it in the demise of the natural world, which we likewise subject to the demand for unification by declaring it part of our tool kit, our construction project. But to love and to know that this love, if it is truly love, is not our salvation, but a contribution we owe to aliveness—this is a bitter lesson, a crossroads. Without the natural world, within which we are loved and thereby placed in proper balance, we cannot follow this road to its end. Without the natural world, which gives to us the balance that we cannot give ourselves, we are ruined.

The thought of the southern midday

Camus gave his vision of a "middle way" a vivid name. It shows the extent to which such an attitude depends on the meaning it receives from our sensuous existence in community with other bodies in an enlivened world. Camus conceptualized his political ecology in opposition to all forms of abstraction that condemn the body and its contradictory needs, that believe one can do away with the reality of pain and of ecstasy. He called his concept *"la pensée de midi"*—the "thought of the southern midday."[7] In French, "midi" means both midday and the Mediterranean region. The Mediterranean, the Algerian coast on the southern fringe of the Mediterranean, is Camus's homeland, the place of his first vivid childhood impressions of existence. "Midi"—the southern midday: The word calls to mind the crackling stillness that lies beneath a blazing sky, the all-but-died-out shrill of the cicadas, the transitory moment of balance in which the world seems to come almost to a complete standstill.

Camus's "la pensée de midi" recalls the poem "To Loll at Midday," written three decades earlier by the Italian Nobel prize winner Eugenio Montale. It is dedicated to the Ligurian midday that, in the parched muteness of light and haze, mixes together the frenzy and the bitterness of life. Montale begins with a description that recalls the scorched life sometimes apparent on bright middays in my hillside meadow in Varese beside the ancient cemetery wall. He writes:

> *To laze at noon, pale and thoughtful,*
> *by a blazing garden wall; to listen,*
> *in brambles and brake, to blackbirds*
> *scolding, the snake's rustle.*

Montale eventually concludes with an acceptance of the boundary:

> *And then, walking out, dazed with light,*
> *to sense with sad wonder*
> *how all of life and its hard travail*
> *is in this trudging along a wall spiked*
> *with jagged shards of broken bottles.*[8]

Montale's poem illustrates the problem of existence and also contains an answer to it. It is an answer like the blackbird's scolding, the plum trees, the rustle of reptilian scales in the arid grass; and at the same time, it is a rejoinder to all answers like those given by other beings' bodies. It is a utopian answer—no suffering, just acceptance. It is "la pensée de midi," because it names the woes but scorches with beauty; because it is the record of a landscape that ignites our hearts to fight for its existence; because it is a manifestation of the soul that fills each and every one of us, because it is body, body glowing with the lust of the midday's heat.

The "thought of the southern midday" is dedicated to life and all of its pain, as well as all of its creative potential. It is not book knowledge but is a practice of knowing at whose heart lies, and whose heart's desire is, the feeling, vulnerable world. In this knowing, the thinking of the world is coactualized, plaintive and yet pugnacious. Thinking about the world—in which I also participate, a life-form in all of my feelings—becomes a negotiation that seeks to protect the planet and thereby to minimize risk. Abstract thought transforms into a practice, a practice of love for the world. Camus says: "the true life is present in the heart of this dichotomy. Life is this dichotomy itself, [. . .] the extenuating intransigence of moderation."[9] For the philosopher Camus, the value of a concept is only earned when it can be avowed by his lived experience, by the wisdom of his skin. This is the simple way in which he avoids taking sides. In our tangible bodies, we can neither keep ourselves from suffering nor gloss over torment. One's body—its trembling; its soft, bristling hairs—attests to the presence of the Other, who becomes real but no longer expresses a statistically measurable size. The author Iris Radisch astutely recognizes the grace of this fleshly candor: "It is embodied thinking, attested to by experience and an awareness of life, and its means are plastic—images, figures and constellations instead of logic and systematic philosophy [. . .], everything that he [Camus] has not thought through completely on the basis of his own experience for him degenerates into ideology."[10] The perception of reality based on the testimony of our own aliveness offers us a measure for life.

In this time of rampant natural crises, climate change, globalization, and the ever-widening worldwide economic chasm between opulence and common destitution, this way of thinking turns out to be more relevant than ever before. Solidarity—empathy for others' pain—is called for, not as a practice of the political set, but as a practice of flesh and blood existence, as a practice of living in a biosphere whose aliveness is being abolished on every level: in theoretical science, in human self-awareness, and in our actions.

We will find no way out of this misery for as long as we try to *solve* it (as sustainability circles and environmental policy makers all over the world are trying to do). Our only way out is to understand that this misery cannot be made to disappear; it must be endured and transformed. Édouard Glissant, a French poet of Caribbean descent, called such an attitude the "thought of the trembling." In this, Glissant is a successor to Camus and his sensory overcoming of the paradoxes that he introduced into French philosophy. The thinker of the trembling accepts ruptures; they make him into a creator and not a cynic. The thought of the trembling is thereby also a maternal thought—a thought that comforts, rather than clear-cuts, in its acts of creation.

To maintain balance means to create. "Balance" is an artistic concept. It describes creative tension with future potential. Balance is a transient equilibrium within a dynamic that imagines more life and acts on its behalf. Balance aims for an ecological prosperity that attempts to mediate between the needs of its participants without succumbing to the illusion that a complete, unambiguous *win–win* situation is attainable. Abiding a dilemma with style, dignity, and poetry is thus a part of maintaining balance. Maintaining balance—acting in moderation—is synonymous with being totally aware without attempting to exercise control.

Mystics such as Richard Rohr call such an attitude "nondual thinking"[11]: thinking that tries to connect with what is—not by neutralizing opposites but by gaining more reality from them. This path leads from a knowledge of the world to a practice of increasing, protecting, and partaking in its aliveness. Like an ecosystem, this path contains the simultaneity of lightness and darkness without censoring either. Richard Rohr calls this "practicing heaven now." Heaven now—this is not the Nirvana that politicians, esoteric thinkers, and technocrats all fantasize about. This is reality in all of its beautiful complications.

Model of aliveness

The British architect and author Christopher Alexander thinks that all forms of aesthetic beauty are really nothing other than various solutions to bringing the unavoidable tension within the process of being alive into dynamic equilibrium. Alexander charts a matrix of life gestures that each represent an artistic archetype and also depict a mediation between the whole and the isolated individual. A point in the corner, a spiral, the transition from light to dark—each of these has the possibility to create a center by producing a separation out a fundamental polarity.[12] Each separation reveals a transition between two deeply interconnected poles. The snail in a mollusk shell, the branching of a tree—these are each forms of negotiation between being and not being, light and shadow, within which the possibility for future growth condenses. They are simultaneously organized and open. They are a home and its potential expansion in one.

It is conceivable that we human beings (but also other life-forms) might recognize such gestures instinctively as the very things that also compose us, symbolically denoting that unconscious rhythm of ebb and flow, construction and destruction, that keeps our biological body alive. The notable thing about Alexander's gestures is that they each contain both poles of reality—light and dark, the whole plane and the individual point, for example—but these poles appear in equilibrium, such that each pole comments on the other, tempts it out of hiding, and so reveals itself to be an imagination of its opposite—not a reflection, but a transformation.

One can even recognize this tension between poles through natural history. From the Big Bang to the future contraction of the universe in a few billion years, the world (historically and spatially—two facets of the same thing) is a gradient between absolute individuation and absolute totality. Every act—historical, spatial, existential—is a balancing act between these two poles, a balance that promises the future. The entire experience of life can be understood as a phenomenon (a vortex, a color loop, an eddy) that establishes balance between these two poles (the Big Bang and heat death), a periodic equilibrium in which being is transformed by nothingness, and nothingness by being.[13]

Both poles must be contained in the forms of this equilibrium such that one does not consume the other; they must yield to a creative tension that perpetually establishes the starting point ("latent centers," as Christopher Alexander calls them) for a new creative mediation of the two polar opposites. The forms in which the natural world expresses itself create something like a waterfall of diverse sensuous forms (which is to say, forms that are in some way connected with life), infinitely extended across time and space. Time and space are themselves aspects of this diversification, this creative tension.

If we follow Alexander in this line of thinking, then beauty becomes synonymous with life—a life that is stretched between irreconcilable opposites and seeks an equilibrium, a state of tension in which the opposites are preserved and can coexist without being dissolved and without being overtaken by one another.

Metaphysics in the mood of loss

Anything promising a "clean" solution by favoring only one side of a contradiction will always be stale and insipid. All forms of hero valorization that neglect to mention that a hero must be ready to accept the possibility of his or her own death first and foremost prolong the siren song that has been the downfall of our Western civilization since antiquity—the siren song suggesting that there might still be some way out (decorated by the trappings of fame and fortune). In all disciplines and all cultural epochs—Renaissance science, Enlightenment, Liberalism, Socialism, ecological Romanticism—sirens have always sung of utopian salvations. As a result, we always run the risk of throwing the baby out with the bathwater. Even in the 1980s, Rudolf Bahro, an ecological intellectual with discerning mind and feelings, longed for a "promised land of ecological salvation"—an absurd formulation, where the Lucifer of the right path is already lying in wait, a Robespierre with aspirations of Demeter and a body of formative powers.

The difficult task for our imaginations is to consider the necessity of death deeply enough that through it we begin to grasp the current wave of planetary devastation. This sixth wave of extinction, which is happening to planetary life right now, is unnecessary and darkly demonic, and at the same time, it might be unavoidable. To say it is unavoidable also means that it will not suddenly happen tomorrow, but has been happening for some time. The ecopsychologist Paul Shepard reminds us that this natural devastation began to become avalanche-like at least a century ago. It is not five minutes to twelve; it is three in the afternoon. To carry on believing that our damnation might just be starting tomorrow is part of this catastrophe. And as with everything, this belief comes from the fact that we are not in a position to see reality.

At the same time, this tragic recognition does not give us license to simply have fun from now on. To get out there and buy an SUV with your remaining credit. To eat some factory-farmed chicken with disposable chopsticks made of wood from the Gabonese rainforest. It means moving toward a recognition that the world is a genuinely tragic place, and that the tragedy of it must be endured, because it cannot be shut down without simultaneously shutting down all of creation. Or rather: not endured, but transformed into aliveness.

This is the great task before us. It results from the increasing precariousness of the modern project of saving the world through technology. But at the same time, it is the task that has always been imposed on every human being, indeed on every life-form. Nothing new, really—just greater than ever before. We can develop here a metaphysics in the mood of loss. This unsparing, pitiless perspective is not a form of nihilism, but rather its antidote.

Perhaps the appalling *grandeur* of our murderous path through this alleged "tunnel of economic necessity into daylight,"[14] and all of the monstrosities of darkness that this path brings to our biosphere, is necessary for us to gain insight into the task before us. A metaphysics in the mood of loss understands this time of dramatic ecological turmoil differently than we are used to: not as a call to swing the rudder at the last minute and finally change, but as an opportunity to accept the unalterable tragedy of existence, for which there is no *solution*. To understand that it is the tragedy of all existence. To grasp that for precisely this reason, we should long ago have sought an existence tempered by moderation in living community with others.

A metaphysics in the mood of loss no longer tries to make the world more efficient. If it did, it would be playing into the hands of the same solution-oriented, economic, and technological thinking that got us into this situation in the first place. A metaphysics in the mood of loss limits itself to mourning pain rather than repairing it. It contents itself with feeling, not with fixing, and waits for what is felt to be real and for needing to be really done, open to every solution, confident in life's desire to heal. It no longer seeks desperately to separate the light from the darkness. But it does endeavor constantly to create more space for the light. In so doing, a metaphysics in the mood of loss strives not for sustainability (which is just another word for technological solution), but for enlivenment (in which there are no solutions, only the creative productivity of life). In our current ecological drama, a metaphysics in the mood of loss recognizes the failure endemic to all creation. It knows that this failure is inevitable. It mourns it without offering an infallible alternative. And to its last breath, it fights to transform this failure into new forms of aliveness.

The middle way:
the emergence of compassion from failure

Francisco Varela powerfully described this expanded perspective on darkness in one of his final essays.[15] He had every occasion to broaden his views in this way, for in the final years of his life, he had an extended back-and-forth with death. Suffering from liver cancer, he received a new, donated liver, completely foreign to his own body, and was very clearly shown the ambiguity of the seemingly firm boundaries of the self—how it is not altogether clear and well defined, what belongs and what does not, and how ill advised it is to come to grips with life based on claims of absoluteness.

At the end of his life, the thinker, in his whole embodied existence, followed the trail of a tragic experience through which he came to feel that there are truly no divisions between individuated subjects, just places where the world folds in on itself—"intimate distances" of varying sizes, but neither identity nor division.[16] The biophilosopher portrayed his own dying with the attitude that he had made his own in life: that of the meditator who learns to consider even his own feelings, his own suffering as objective occurrences in the world and to not imprison himself in his own needs.

If you can manage to see yourself from the outside and accept yourself as a vulnerable organism within the biosphere, rather than focusing on the narrowness of your own needs and considering yourself the center of reality, you will be amazed at how the world positively overflows with suffering beings, each with needs and appetites of their own. Something odd occurs when one has that experience: Suddenly, the soul is filled with deep compassion for the world. This is the only attitude adequate to the task of scaling the dark mountain of the ecological catastrophe we face: without any illusion that it is possible to change human beings based on the strength of better insights.[17] But with full-fledged empathy for this creation, among which things are so hard precisely *because* it is creation. And with equally strong gratitude for this creation beneath a graciously warm sun and its affirmative light: where we have already been given everything we need to live, before we ever thought to desire it.

Varela had described this perspective in detail in his book (written together with philosopher Evan Thompson and cognitive researcher Eleanor Rosch) *The Embodied Mind*.[18] For Varela, perception was always play with imaginative surplus. This is why it turns out to be a genuinely nontotalitarian act: It is a constant negotiation between two sides, and accordingly, it is always an interaction between light and shadow. In Varela's view, perception assigns equal weight to both an implacable reality and the fleet-footed imagination that alters it. Such a perspective understands the world's becoming as a dialogue between bodies, an ongoing act of erotic touch and erotic transformation.

Here, the ancient Eastern and Western thought traditions of the last two hundred years might perhaps coincide, if we bravely unhinge Western thinking from its desire to dominate and sound for the fainter traditions of connecting with the immanence of living sense, like Camus attempted to do. We know how the technologically successful West will reflexively respond: "To arms! And clean up this mess!" An energetic application of seemingly correct knowledge, the flattening of contradictions, the redemption—even if it requires imprisonment—of those who resist. The classical East responds with insight into reality, but in the traditional spiritual practices this insight often tends toward a withdrawal, a surrender of the self, a refusal to act in order to retain the ability to see. Although the East has approached paradox much more openly than the West, which has been denying it for three thousand years, inaction is not the answer we need. Paradoxically, we need both: action, and the knowledge that action is never enough—indeed, that there is always something absurd about taking action.

Only the vision of a society that engages in this imperfect action, that admits both tragedy and the "astonishing beauty of things," in the words of Robinson Jeffers, will allow us to recognize a genuinely new reality.[19] Only such a vision will be in the position to turn away from the dominant social technologies and economic theories and toward a practice of participation in which no optimum state can be reached, but in which we can constantly struggle to bring about an optimal equilibrium. Only such a vision accepts that we are part of everything and cannot separate ourselves from it through either our acts of destruction or our achievement of excellence.

This variant of the "back to nature" idea does not mean that we have to go back to living in huts. Above all, it means relearning to feel and to perceive, and accepting all of the feelings and perceptions—the terrible and the sublime—as reality. As our inner share of wildness, as our inviolable participation in a pulsing reality. With this attitude, it makes no sense to look for guilty parties. Here, acting in moderation becomes a gesture of immeasurably deep compassion.

— *chapter eight* —

SHARING

We are made to enjoy giving.
MARSHALL ROSENBERG[1]

"**S**tronzo." *Asshole*, my friend Luciano says to me, and I know that I have finally arrived. Luciano looks at me, tough, amused, and thankful.

"Stronzo," he repeats, awkwardly, tenderly. The words are my victory. A suddenly swelling wave of gratitude engulfs me. I am there, alive, in the midst of reality.

Sure, I had only made it possible for Luciano to give me this gift, with which I was finally welcomed into the present in Italy. There, at the scuffed metal of the counter in the Bar Sport in Varese Ligure. I had done Luciano the favor of buying him a sandwich, that was all. And an espresso afterward. After three months of living in Italy.

Three months of having a tough time of it with Luciano.

In scenes like this:

"I'm buying." "No, out of the question." The money is already lying on the bar. I'm too late. Too slow.

And this:

"Let's at least split it 50-50." A distant smile. The euros already clinking in the waiter's hand.

Or this:

"You can't pay for the whole dinner. That's no good. What did it cost? Tell me, please!" No answer.

Or this:

I go to the bathroom, thinking, the glasses have only just been ordered, they're basically full. So nothing can happen. I'll have my chance afterward. When it's time to part, I hurry to the bar before Luciano, fast as an arrow, like a hunter on the attack, sure of victory, but the bill is already paid.

It is a sardonic game that I always lose. Always. My purse remains full and my shame grows. Soon I won't even be able to go drink a small coffee with Luciano.

"That's a game for rich people," says Alessandro, another friend who, unlike Luciano, isn't from Naples but from Padova, the backcountry of the traders' republic of Venice.

"No," I reply. "It is a game for poor people. Even the poorest of the poor can play it."

"Stronzo."

This time I thought ahead. I instructed Walter, the host behind the counter. An elegant setup. I gave Walter carte blanche: "Whatever we eat and drink, I am going to pay you afterward. No matter what, no matter how much he insists, you tell Luciano that the bill has already been settled." Of course Walter played along, an innocent smile on his face; he was happy to have this opportunity—he is always happy when he does somebody a favor.

In front of the Bar Sport, we shook with laughter. Luciano had given me the gift of allowing me to give him something.

It cost him nothing. He had saved money, in point of fact. He had just saved money. I, on the other hand, have been enriched.

Outside, the swifts circle the old castle tower. The thick trees on the hill behind, the sweet chestnuts, oaks, ashes, and cherries glow as though warmed from within, orange in the late afternoon sunlight. I would have to describe this light . . . no, I would have to describe light, in general, as transforming itself within my soul into a feeling of radiance. For the first time in months, I can imagine trying to do so. The shadows are short, the irregular granite paving emanating heat. From the bakery comes the smell of fresh white bread, of garlic, and of sheet cakes with mangold.

It is a game, and I won by voluntarily losing.

I have offered a sacrifice and: I am alive.

To gift is to give life

During my years in Italy, when I had the little apartment in my Ligurian town, Luciano taught me about gifting. He thereby introduced me to the practical aspect of the "thought of the southern midday," and he still reminds me of it each time I see him. He doesn't make it easy for me to train in this practice: If I want to give him something, I have to outwit him. And how am I to do that with him, a Neapolitan *signore*?

Superficially, I might even be relieved: Luciano's generosity helps me save money. This is the perspective that my other friend adopted when I tried to infect him with the lavish thought of the southern midday. (I paid for him several times in a row before he even noticed. So it was a game for rich people after all.)

I learned from Luciano nothing more than the fact that I could give life and that this gifting imparts to me a feeling of being alive myself. There was a deep satisfaction in giving, a happiness that befitted the buzzing sunlight, the levity of warm air at midday, the weightless joy that streams from the Ligurian hills on beautiful early summer days, when everything was given and the whole world belonged to me. Our tricky struggle over who would put the two euros for the two espressos on Walter's bar was, in truth, a game full of possibilities to give life to the other: an exercise to learn that the path to I leads through You.

When I pass on something that I truly lack later, that makes me more enlivened. This is the doctrine of generosity. It makes me happy that another's joy has cost me something. In order to give life, what I give must be a part of my life. Something whose lack I feel, whose loss limits my own freedom.

The gift must be an act of externalization if it is to engender aliveness. Pain thereby becomes part of giving to others. This, too, is part of the thought of reality in the mood of loss: For as long as we are alive, we have the opportunity to share this life and its resources. There is a deep connection between unconditional giving and the unavoidable failure that will one day extinguish every act of creation, the results of every action, every creature, every world improvement. Giving means both retaining and relinquishing balance. In giving, failure has already been sublated.

Failure is wastefulness. Overspending. Giving to the point of excess is something like an anticipation of the fact that individuals can only be productive because they do not belong to anyone, including themselves. Giving is a gesture that visibly places the aliveness of the whole before that of oneself—something like the attitude of motherliness toward Being itself.

Such reflections come more easily in summery Italy than in northern climes. This attitude can be more easily tried when one lives under a buzzing light. There, one has the direct sensory experience that generosity or even extravagance is the proper attitude toward life. When the light bathes your body on a summer morning after you open the shutters, the overwhelming recognition of the skin, the happy assurance of the retina, is that it has received a gift. Camus's whole philosophy originated in the southern sun and its munificence. For him it was not only compensation for the hardships of an impoverished childhood, but it also embodied the principle that made it possible to imagine a reconciliation.

We receive light as a gift. Light is the epitome of that which is given free of charge and without ulterior motives. It is also derived from an act that is completely void of intention: The sun bestows its warmth and wastes itself in the process. This is why Camus was so shocked after he left Algeria and arrived in Paris. He noted that it seemed to him as though he only then learned what true poverty was—there, trapped beneath a gray and rainy sky—even though he had grown up in some of the most modest conditions imaginable. The southern poor were at least recipients of the light and thereby connected with the foundational power of existence.

It is a matter of understanding selflessness in action as a central principle of aliveness and thus as a principle for how the self produces itself. Feeling a fiendish joy in outwitting Luciano so that I could cover his costs, I pursue selflessness as an ecological principle. I experience it as a principle of successful relationships within an interwoven network of distinct identities. The happiness felt beneath a sun that gives without design or intention is the tangible marker of such an experience. Our bodies allow us to share in it.

An ecology of the gift

In the natural world, everything is gifted. From the sun's warmth to the sustenance that it donates to us, from the possibilities of deep self-understanding to the joy that results from it, nature's gifts are exuded as unconditional offerings. It is characteristic of the natural world to be *life*. We receive everything living free of charge. The vivacity that we experience in the old, densely interwoven network of relationship is an offering without expectation of return: an expression of life that enlivens.

We should absolutely understand the connection between the self-externalization that is part of all living relationships and the giftedness of the living nature that is everywhere apparent. Only when we comprehend that the circulation of this offering is a central precondition of ecological flourishing can we set aside the avaricious attitude that is currently so pervasive. Only when we grasp that the happiness that "returns to our own hearts" is at its core an echo of a state of ecological balance can we truly savor the act of giving without fearing for our own substance. For there is no such substance. There is only the transformation of diverse subjects into ever-new forms of mutual recognition, the creative imagination of personal futurity. Even our bodies do not truly belong to us but slip constantly through our fingers. Even our metabolism follows a central ecological principle—that externalization without intention is the only thing that creates identity. "The person who gives," says the philosopher Lewis Hyde, "is a person willing to abandon control [...] and participate disinterestedly in a circulation he does not control but which nonetheless supports his life."[2]

More than almost anyone else, Lewis Hyde has investigated the importance of giving deliberately (as opposed to the strategically employed service) to the fruitfulness of creative processes. Hyde sees that there is a mysterious connection between giving, life, and our creative power. He talks about an erotic of the gift and writes: "as gift exchange is an erotic commerce, joining self and other, so the gifted state is an erotic state: in it we are sensible of, and participate in, the underlying unity of things. [. . .] Gifts are the agents of that organic cohesion we perceive as liveliness."[3]

With this, Hyde placed a crucial correlation center stage: The degree to which our actions or works are productive has something to do with how much they engender aliveness. According to Hyde, this can only happen when one's own actions are not solely directed toward safeguarding a vulnerable ego. When they are not subject to a dictate of performance or a compulsion to control, but seek to pass along the fullness that one has received oneself—and thereby to give out something valuable, indeed something essential for life. If giving is a central ecological dimension, actually the center of any ecology, it follows that we must make it also into a cornerstone of a culture of life. But what would this look like?

Hyde found few examples for a culture of the gift in our current civilization, but many more in archaic societies. It seems that the more connected to nature a certain society is, the more it defines itself through a gift exchange with the rest of the world. The culture of the gift is always based on an understanding of nature as the ultimate source of an offering dispersed without any reason. Life is a gift, not a reward won in the war of all against all, as we have learned most recently following the triumph of vulgar Darwinism.

Many people who live in balance with the natural world feel that they receive gifts from it, and also feel prompted to give to it. This is the core of the archaic recipe for the protection of natural "resources." That which serves life is understood as a gift, not as goods that are lying there to be used for one's own advantage, without giving back anything in return. A gift offers life and the joy of gratitude for the gift of life. This joy longs to reciprocate the gift, to spend something for the natural world. Many archaic peoples therefore make it a practice to give something real back to the wilderness whose nonhuman inhabitants nourish human beings with their bodies. Hyde reports about a Maori tribe whose members regularly carry parts of the quarry from their hunt and the bounty of their harvest into the woods in order to symbolically nurture the productive power of these other beings. There, the food breaks down, consumed by animals and transformed by mushrooms, and it does, in fact, reenter the cycle of becoming and dissipating.[4]

The members of such a culture, Hyde thinks, act in such a way as to give something back to the beings that offer them life. They know that the "circle of the gift" must not be broken if the nourishing powers of the natural world are not to run dry.[5] Hyde sees that acting in this way is not just quaint, cultural folklore, but accords with a deep ecological insight. But this insight actually has nothing to do with trying to create an advantage for oneself. It doesn't mean protecting one's own resources the way that an investor protects his cash reserves in order to wait for a strategic gain. An offering that one gives, even though it was badly needed to satisfy one's own hunger, accords with an insight into reality. According to this insight, nothing living can be possessed. It obliges one to offer up a part of one's own life comforts in order to strengthen the aliveness of all others. This is neither based on the calculus that this is the way to get ahead nor on the obligation to serve the greater whole, but it embodies the knowledge that nothing is entirely one's own. Life force cannot be monopolized and hoarded. A share of it simply falls to you. This insight expresses itself most strongly in the need to repeat this cosmic gesture and to allocate the energy of that life force unto others.

Those who belong to such a culture of ecological unreservedness do not delude themselves in their picture of the natural world. For the natural world also does not follow a model of well-controlled frugality. Obsessed by the drive for efficiency and improved performance, we err by thinking that all relationships in the living realm are the result of uncompromising competition and keenly calculated and optimized expenditures and that they all are in service of evolutionary progress.

The mainstream of evolutionary thinking begins with the idea that all structures and manners of relating within the living realm must be the result of a kind of natural, self-organizing cost–benefit analysis. Competition has long been the undisputed dogma of the biological imagination. Mainstream evolutionary thinking aligns its analysis with these premises, instead of first observing the world and then drawing conclusions. It imposes onto reality the afflictions of a society in which uncalculated joy in life is at risk of being suffocated within a network of greed and oppression.

Being eaten and being gifted

I have discussed in detail the dangerous errors of our view of the natural world as an outsized McKinseyian headquarters in the books *Biokapital* (2008) and *Enlivenment* (2013).[6] In the following I will therefore describe my memory of an experience that allowed me to grasp, with my body alone, the extent to which complete and total extravagance is a precondition of aliveness.

It was June, and the fireflies were coming out. On a quiet evening, my dog and I climbed the narrow path through the chest-high grasses of the meadows into the hills behind my Ligurian village. The blades of grass vibrated quietly in the night. At the end of the stairs that led to the cemetery, I looked up, and it seemed to me as though I dove straight into the stars, headfirst. The fireflies filled the air above the meadow and the trees. They flew like little stars, dancing zealously through space and flashing off and on. Their light mixed with the silver points of the heavenly bodies in the darkness of space, blurring the line between star and glowing insect. It was as though the stars sank into the grass and the smoldering insects ascended into the universe. For a brief second, I had stretched my head out into a living cosmos.

What I saw was a festival of reproducing and of being devoured. I looked up between the stars and participated in a dance in which the participants ecstatically offered their bodies to others for use. The glowing insects danced through the night to find a mate and were consumed by late-flying birds and bats in the process. Everything gave itself away in the truest sense of the word, without wanting to, without thinking, without even being capable of having a thought about it. And it was no accident that it unfolded in this way—it was not some regrettably still-unperfected ideal.

This glittering fullness over the meadows could only be realized in the nocturnal dance of eating and being eaten. The gift was needed in order for the whole to exist. It seemed to me that the gift, the selfless expenditure, integrated the individual and the surrounding whole. The received gift belongs completely to me, but it belongs to me only so that it can be given again, passed on to someone else. Moreover, it only belongs completely to me once I pass it on, just as the stuff of my body belongs completely to me while at the same being completely unavailable to me.

Ecologically, it is essential for meadows to be grazed (or mowed) in order to maintain themselves. They must be wasted in order to prosper. Perhaps 0.02 percent of the seeds produced by the grasses will eventually turn into adult plants. Perhaps just as few of the eggs hidden by the female fireflies will develop into new, smoldering insects next spring. But the continuing stability of this landscape's existence owes everything to this impossible waste that nourishes all beings, thereby contributing to their actions that will allow the meadow to go on wasting itself. It was nights like this one that caused me to trust in the generosity of life, to gift my trust to aliveness as such. The tumescent grass that I had not summoned, the flashing points of the insects' light—these things made me into an unfailing optimist. They transformed me into someone who sees the gifts that he otherwise ignored, into someone who trusts.

We must get used to the idea that the opposite of efficient frugality is dominant within the natural world: immoderate, senseless waste. Even those beings that we often refer to as "efficient hunters" in early-evening TV specials—the big predators like lions, pumas, and wolves—are shockingly inefficient. For warm-blooded creatures, including us, more than nine-tenths of the energy we take from our nourishment is lost to maintaining our body temperature alone. The energy audit there is as disastrous as in one of those prefabricated houses from the 1950s that has not been thermally insulated.

Just as the energy usage of these great hunters is not a model of efficiency but its opposite, they are also not truly well adapted in other ways. They are anarchists of self-expression, dandies among the animals who can afford any sort of extravagance. The Prague-based biologist Filip Jaroš recently found through comparative studies that the striking coats of jaguars, leopards, and tigers in no way blend in with their surroundings, as zoologists had assumed to be the case—and as generations of schoolchildren had been taught in their biology classes. Stripes and spots are not camouflage but are actually a painted warning. If you shine yellow among the green shrubbery and are wearing a striking pattern on top, you stand out. But the great cats are so strong and so fast that they catch their prey regardless. Their appearance is therefore not a functional tool, but a surplus, an excess. The patterning of the predatory cats is a gift given by life to itself.

Considering this way in which the biosphere works, ocelots and cheetahs are not exceptions but are expressions of the biosphere's priorities. In other words: The natural world is only no-nonsense and functional when it has to be. A pork tapeworm has no extremities because they would interfere with its survival. But when arabesques do not get in the way of subsistence—as is true for the tapeworm's distant relatives, which are adrift as plankton in the turquoise of the open seas—forms develop that are like stylized works of art in their fantastical exuberance. The drive to exhaust all possibilities of design and construction is not only allowed but seemingly corresponds to a secret desire in reality for expression and new form. And this abundance of forms is a continual gift.

In the natural world, the most complicated, most valuable things are given away without a thought. Its abundance is not conducive to investment but is an act of trust that is constantly redeemed in a grandeur that seems senseless to frugal capitalists like us. How are we to understand, for example, that a wolf—this marvel of presence, tenacity, tenderness, and physiological precision—has lived its full life after just seven years; that an animal of such perfection falls into ruin after such a short time? How can we dare, as Cormac McCarthy writes, "to hold what cannot be held [...] at once terrible and of a great beauty, like flowers that feed on flesh"?[7] How can we comprehend the boundlessly deep miracle of the mayfly, on the surface of the water no less perfect than the wolf, flawlessly and meticulously elaborated, yet nothing more than an empty shell after but a few days? And what does the may bug have to tell us, that beetle that enjoys a few early summer days after five endless years of larval existence below the earth? The single book, completed after fifty years of work?

All of this shows that the self, the delimited life of the individual, and its accumulated treasures cannot be the standard. The self is not the pinnacle of worldly affairs. That for which the incomprehensible filigree of the self is sacrificed must be infinitely more filigreed, filled with infinitely more self. Clearly, the wasting of so much preciousness is not at all significant to the greater whole—but not like a general, for whom the slaughter of a few hundred soldiers is more or less insignificant; rather like a wave that is indifferent to sand that it shifts with its driving surf, like a crystal that is indifferent to the position of the individual silicon molecules that it absorbs into itself. The sacrifice to this completeness, the most light-footed giving away of this *focusing of the world into a powerful I*, is perhaps precisely what makes the gift complete. It is not a question of somehow taking advantage of the time you have and thereby achieving something. On the contrary, it is necessary to choose what you consider important and to completely devote to it beyond the self, so as to make the length of time that remains insignificant.

Gary Snyder observes that we should understand "the play of the real world, with all its suffering, not in simple terms of 'nature red in tooth and claw' but through the celebration of the gift-exchange quality of our give-and-take. 'What a big potlatch we are all members of!' To acknowledge that each of us at the table will eventually be part of the meal is not just being 'realistic.' It is allowing the sacred to enter and accepting the sacramental aspect of our shaky and temporal personal being."[8]

The food chain is the extreme example of a relationship through contact, a relationship through transformation. On the basis of this touch, on the basis of the possibility to enter into a greater unity that this contact affords, some cultures did not actually consider being eaten as symbolic of dying, but of sexual union. A symbol for a kind of union that is also a transformation, the result of which condenses into new life and new individuality.

Freedom needs the greatest gift

The food chain is part of a creative process without which the ecosystem would collapse. For each individual, it is the only basis of personal preservation. We humans are also alive only because we are part of the planetary food chain. The food chain is a clear negation of distinct individuality. Not right away, but one day we, too, will become food for someone else. The stuff of our bodies will thereby transform into the bodies of other beings. The creative freedom of the natural world is only possible because it constantly devours that which is elsewhere produced.

Freedom has a price. And this price is death. Without death—which pulls a creature back in at some point—no self-preservation, no self-will, no insistence of one's own position, no creation. Therein lies the deep logic of what Gershom Scholem once hesitantly suggested: that creation cannot know perfection. The freedom of creation, the creative autonomy that speaks from every blade of grass, only manifests if an individual stops taking first for herself and instead totally risks herself for it. Freedom is not a possession but a form of nakedness. To be able to receive it means to give up all claims of anxious self-preservation, all withholding from creative exchange. It means giving in to imperfection and to the suffering that goes with it. Freedom—this basic power of every being over the stuff of its body, this furtive, sought-after goal of a natural history based in ever-increasing autonomy—is only gifted to us if we are constantly willing to give our life. It is the dimension of existence that demands life back as a gift. This is the innermost core of the labor of gratitude.

We should not forget this, we inhabitants of societies that apply the label "free and democratic" in every spot imaginable, by which we mean as free as a shopper with a loaded credit card in a mega mart. No, the freedom of living is not money raining from the heavens. It is the opposite of that—and yet it is connected with it. Freedom appears in the actions of the girl who gives away her last shirt to help other freezing people. It is that which demands all of our courage, being alive.

Here, we see clearly the unavoidable connection between creative freedom and failure, the collapse of the project of private identity. Gifting means acting like an ecosystem in which something is constantly made available at the expense of the individual, thereby elevating the resilience of the whole. And at the same time, my giving as a freely chosen act affirms my individual sovereignty to create aliveness.

We can only learn to understand how poetry and the food chain are related through the idea of the gift. In that idea, we comprehend that participating in creative reality means giving something up and receiving something else. In that idea, we can grasp that whenever we venerate the world like a child, it enlivens us. And we love our children by adding something to the world. We love children in order to liberate them, not to possess them. To constantly enlarge their freedom.

Like poetry, like love, like the rapt and agonizing commitment to a collective concern, like a stirring idea or a humorous insight, aliveness is something that increases when we share it. Every true gift is given not just to an individual but to creation itself—it serves to offer that creation more life. Just as every new ecological niche is less the triumph of a "fit" species than a deepening of the relationships within an eco-system—and an intensification of its potential for life. The humanist psychologist Abraham Maslow remarks that the people who have the most alluring effects are those who subordinate their own interests to the elevation of aliveness.

The philosopher and writer Friedrich Schiller had indeed called such a person a "beautiful soul." And Maslow's colleague, Ernest Becker, infers from this: "In the creative genius we see the need to combine the most intense Eros of self-expression with the most complete Agape of self surrender."[9] And, Becker continues, referring to the German psychologist Otto Rank, that only in this way, "only by surrendering to the bigness of nature on the highest, least fetishized level, can man conquer death."[10]

A practice of love follows this creative genius. A practice of love is always an ecological practice. It makes possible an ecology of life-bestowing relationships. It is profoundly creative because it does not create any hindrances for the open-ended unfolding of transformation through relatedness—rather, it nurtures that unfolding even when it means accepting self-sacrifice, one's own (symbolic or actual) death.

Love is the gift wherein the reciprocity of giving is fully realized: My love for you is gifted to me so that I can gift my love *to you*.

The gift of perception

Love is the reason that we long to be near to the natural world: Together with other beings, we can experience this form of life-bestowing reciprocity, insofar as it befalls us as a gift, as something that transpires simply as it is, that happens for no reason, and that we are likewise free to pass along. Being in the natural world means being in a state of relatedness and also grasping the terms of relationships within a creatively enlivened world. It means perceiving and also sensing the basic principles of mutual, living perception. When lived attentively, life never loses its connection to these principles. It is always both action and experience, the experience of being one of countless centers in the universe.

Today, the greatest of catastrophes lies in the fact that we no longer take it for granted that we grow up in a cosmos wherein life is gifted to us. For life *is* gifted, regardless of our relationship to it, just as sunlight is gifted. Not to treat it accordingly amounts to an enslavement of the creative impulse, an act through which all creation comes to an end. Ultimately, we are most deeply threatened by the consequences of this, by the end of creation, and not just by material catastrophes such as climate change or extinction. The first act of resistance to the current menace of life therefore is not eager protection of the bits and pieces that remain, but the confidence that we can be alive if we act as parts of the circle of giving.

In front of the old oak tree in the Grunewald, not far from the Heerstraße S-Bahn stop, the preconceptions of our age are fading—an age that misunderstood reality, because it did not share it with other beings, with those other beings who had no choice but to respect the fact that they do not have technology that offers them bad credit in exchange for their entire future as collateral.

The last bit of snow lingers on the rustling leaves of the forest floor. The oak tree is at least five hundred years old. From its stout, twisted trunk, a few massive limbs reach with contorted gestures into the empty air. Here and there, they have long since rotted. Among the other plants of the forest, the slender pines and birches, this tree stands in silence. It dominates the little private clearing that has formed around it, a crooked figure of geological time, a whole world full of all its contradictions, powerful and broken, softly eroded and solid as stone, damaged and upright, coarse and graceful. Its bark gleams brown and red, the lichens gray and blue, the moss green and gold. The branches that stick out erratically from the limbs indifferently grasp for time, whereas the trunk is nothing other than itself, accepting the steady trickle of the passing minutes, the gentle drip of good and bad along its cracked surface.

A line of cranes divides the sky. Their fluting calls sink through the empty whiteness above the pine fronds. I stroke the burst, cracked trunk of the oak and ask myself what it feels when my skin touches it—could it express it in words? I feel warmth, restrained and velvety hardness, finely delineated structure, calm, symmetry. And then, startled, I perceive the tenderness with which something, someone gives me their attention when I lend my attention to them. And I am flooded with the joyous feeling that every encounter is a communion, an exchange of gifts, a feast.

Every moment of living existence is perforce an indulgence of this gift. Every moment of living perception contains the potential for revealing our existence as a practice of love. When we succeed in slowing down our experience enough to pay attention to our individual senses, such that our feelings and their interaction with the whole of the world become perceptible, something astounding happens: With a sigh of tender gratitude, we understand we have always already been the recipients of a gift.

As the philosophers say, every perception is given to us. The tickle of the delicate drops of a summer rain on the skin awakens and enlivens our pores as though they had freshly emerged from nothingness and makes our skin tingle as though it had just closed over an old wound. Therein lies the deep emotion and untamable stirring that causes poets like the previously cited Gerard Manley Hopkins to express gratitude for things, gratitude for what is. In such experiences, reality expends itself for our senses. The German poet and philosopher of poetic thinking, Rainer Maria Rilke, thought that his only role in life was *to praise*: and that means nothing less than calling reality into life like a child unwrapping his presents beneath the Christmas tree, astonished and incredulous.

The world differentiates itself only in the presence of countless bodies, cells, eyes, buds, wings, and lips, when individuals reciprocally bestow reality upon one another in relationships. This is *interbeing*: generously enabling existence in a web that has been created by all. "If [. . .] we and the beings of the natural world reciprocally create our perceptual capacities through our entwinement with one another," says the psychologist Shierry Weber Nicholsen, "then this cocreation is a reciprocal gift giving."[11] And she goes on to say: "Making ourselves worthy may mean being willing to suffer the catastrophic change initiated by opening our perceptual capacities to the gifts of still deeper perception."[12]

If you pay attention, you will notice that the gift has already been distributed everywhere, and that we need only reach out for it. What befalls us in everyday life, and what we obviously assume to be given, is precisely that: a gift. She offers me a smile. The skin that closes over an injury all by itself is a gift. All processes of self-organization—which is to say, all physical and biochemical processes, those that the systems' researcher Stuart Kauffman has described as "order for free"—are gratis, are gifts. New capacities develop in systems only because they become more complex over time. Those capacities are awarded to the systems from nothingness.

The world, it seems, longs to receive gifts and to offer new things itself. A new cosmos opens up here; the world of an ethics of the gift. Every perception is a gift to those who produce it. The "flesh of the world," in which the French thinker Merleau-Ponty saw all relationships embedded, is not just made up of acts of mutual perception; it entails the reciprocal offering of life through these perceptions, which always involve contact in the senses. Every act of perception, every trembling of skin, every flash of a quantum of light in the apparatus of a nerve cell is a mutual act of give-and-take. The "flesh of the world" is a network of reciprocal grace, an unfathomably interwoven practice of giving. Everything that offers itself as visible and tactile parts of the world calls for a gift in return. The voice of poetry resounds in this call.

A deity that gives extravagantly

The philosopher Hans Jonas likewise has thought about the gifted quality of the world. And he also approached it from an unexpected angle, from that of a victim. A victim who experienced an incomprehensible monstrosity. Jonas tried to understand how the traditional Jewish concept of a benevolent and all-powerful God could be reconciled with the catastrophe of the Holocaust. Jonas thereby posed anew the old question of the possibility of God given the existence of evil, and he posed it at a point in history when the idea of God—and of poetry also, according to the influential thinker Theodor W. Adorno—had gone finally bankrupt. But despite such verdicts, the question remains unanswered and is posed again every second. It is the same mystery that we face when we consider the connection between the poetic abundance of the Ligurian meadow and the fact that the technical term for this abundance is the "food chain."

Jonas did not offer theological comfort in his answer. He accepted all of the incomprehensible pain of what had occurred and also tried to anchor it in a dynamic of a fundamentally good world. For this, Jonas invented the idea of a "cosmic daring."[13] To do so, he distanced himself from the model of a God who stands above the world and instead embedded God into the things themselves, which thereby became His own becoming and which He has been powerless to control since the very beginning. In order to become real at all, Jonas suggested, divinity had no choice but to transform itself into the world, to extend itself into the history of things over which it had no control, because it was those same things—but in the outside of them. In Jonas's telling, God is not outside the world, nor in the world, but *is itself* the world and all of its possibilities, including the most wretched. In order to return to the divine, it is necessary to accept this exorbitant gift, the gift of being inside God, of being God oneself through one's own aliveness, to pass it on, and thereby bestow life.

The "cosmogonic myth," as Jonas carefully calls his notion, depicts the divine as an outrageous gift that cannot actually become real if it is not accepted and made one's own. Only by taking it on ourselves can we have a share in it. Thus, the divine has two sides: It is the desire, felt by all, for individuation within abundance and the courage to take on the "task of gratitude" and to transform this desire appropriately into a reality that welcomes life.

We also feel this impulse, having ourselves come from the matter that emerged during the Big Bang, and being ourselves children of a chain of living cells that dates to the beginning of the biosphere. Perhaps it is there that the undeniable feeling of aliveness becomes evident, the one connected with our interest, as vulnerable and imaginative bodies, in sharing a world with other vulnerable and imaginative bodies. In the wish to be more real is hidden the desire of the whole divine undertaking.

It is this wish that we also share, that we also feel within ourselves. The mystic Thomas Merton calls this experience the *"point vierge"* (the virgin point) of pure aliveness that wishes for life for the world, not for one's own life at the cost of others; it desires that life exist within reality.[14]

The virgin point allows us to recognize whether preference has been given to aliveness or to efficiency and control within a family or a landscape. It allows us to measure whether we are truly alive ourselves, or merely functioning. It is the quiet voice of truth that Susan Forward speaks about—that voice that always understands whether we are living up to the principles of living exchange, for ourselves or for those who follow us. It is the place within us that ensures that we can only truly be ourselves when we allow others—other people, our children, the meadow, the forest, the old tree, the mole in the garden—their place within the web of reciprocity. The fertile ground of our affection for the created world, and the joy and lightness we feel whenever we are able to offer something and bring joy to the recipient. This place does not belong to us but to the life within us. It, too, gets larger whenever we waste it.

According to this view, the traditional religious notion "God is grace" would mean: The creative power gives of itself without moderation, without reservation, divesting itself to the point of mortal injury. It would mean that the godhead distributes herself without any restriction, without any goals, and without any expectation, just as we, too, are given to the world and thereby to ourselves. It would mean that following this life-bestowing gift, we are called to take on the labor of gratitude so that we might thereby be able to give something back. It would also mean that whenever we receive the gift of existence and all of its possible manifestations, from the crassest destruction to the most magnanimous extensions of self, we are needed to take on the position once thought to be reserved for the creator. If life does not engage in the labor of aliveness, the godhead is not only helpless but cannot come into being. We are the ones that *can* assume "the divine cause"[15] (as Hans Jonas says) by passing on the gift that is the divine itself.

We pass it on by bringing forth life, by enlivening others and making them real. The animals and plants pass it on by promoting the inevitable exchanges of ecological processes in a state of unthinking abandon. The natural world is nothing other than this: the circulation of this one great gift. Hans Jonas is in accord with the mystic Simone Weil on this point about an ecology of the gift: "Love is not a state of soul; it is a direction. It is, so to speak, the axis of two poles, one of which is matter, the other God [...] God can only be present in creation under the form of absence [...] This world, in so far as it is completely empty of God, is God himself."[16]

The ecology of the gift enables a practice of love. Love as a practice demands that one carry on and thereby understand the act of creative giving, by constantly producing oneself anew as part of this living fabric. The ecological side of this understanding includes my ability to understand that by existing with my senses in constant relationship, by not judging, by not concealing my needs, I already carry the entire creative biosphere within myself. Because of this experience, Richard Rohr says, "your life is not about you, but you are about Life."[17]

Jonas referred to this potential to be able to assume and to pass on the self-expenditure of the creative force as a world-creating "cosmogonic Eros," and contrasted it with a "cosmogonic logos," which rational science has tried to employ with maximal efficiency for the past few centuries.[18] But the Eros of a reality that longs for constant creation, for unfolding, experience, and self-experience, is precisely the thing that we cannot forget, for it does more than determine the proportions of our connections according to measure and number—it guarantees the subliminal drive for individuality and the desire for unification. We cannot wait for it—we must make it into a central concern. Only then have we truly understood Eros. Only then will it make us beautiful and alluring, beautiful like every animal that follows the Eros of creative necessity with every fiber of its muscles and every hair of its pelt, never questioning its own conduct. Only we can nourish this Eros for ourselves. No one else, no distant God on a cloud will do it for us. The Eros of the living network is given to us, to us alone.

A font of the imagination

The grosbeak is there again. He is sitting on one of the limbs of the pale green hawthorn in the snow. The bird staggers a little amid the streaks of white that have been deposited in the forks of the branches. Because of his massy, stout body, I thought that the animal was a jay at first. A little above the branch, a pair of pied woodpeckers are carving up a spot bare of bark. The animals rip the bark up and leave behind the traces of their lives in the wasting wood. The tiny larvae and little invertebrates that hide within it feed them.

The tree is beginning to break down, to die off, and thereby to become a gift for other beings. It is beginning to be less itself as a sovereign individual that decides only for itself. And at the same time, it is beginning to become even more itself by becoming lively, a hub of activity, a gift that enriches others. The hawthorn is relinquishing itself and becoming a center of many existences, a fervent font of the imagination that circles in the ice-cold air about this middle, which is gradually becoming ever more empty.

In the spring, an attentive gardener will see the traces left by the woodpeckers, will discover the tree's self-expropriation, and will intervene without thinking much of it. He will trim the "dead wood," will paint the wounds with grafting wax, will perhaps call in help, and he and his colleagues will fell the "sick" tree and carry it away. Then everything will be okay for them again.

No one will notice the message visible in that bare hawthorn, the insight that is revealed by its crooked, icy-gray trunk, bedecked by lichen and rime, grosbeak and woodpecker. It tells us: Death is the way in which the living make an offering so that more life emerges. The tree will be repaired or carried off, for no one is supposed to notice this message. Its contents remind us that our own flourishing is indebted to a gesture that is painful because it wastes on others the very thing that we fought so hard to gain.

Don't worry. The gardener will arrange everything so that the plants can be admired as radiant victors, each standing completely alone, triumphant over all the others.

— chapter nine —

THE HEAVENS, NOW

*I . . . saw the circulation of my dark blood, saw the coils
and springs of love and the alterations of death, saw
the Aleph from everywhere at once, saw the earth in
the Aleph, [. . .] saw my face and my viscera, saw your
face, and I felt dizzy, and I wept, because my eyes had
seen that secret, hypothetical object whose name has been
usurped by men but which no man has ever truly looked
upon: the inconceivable universe.*

JORGE LUIS BORGES, "THE ALEPH"[1]

The heavens kept their promise from the day before. Today they curve above us in transparent blue. And it is there again, that first sky of childhood, the sky of that one brilliant day that stands for all of the long-since-bygone hours of a carefree perception of happiness. Back then, it became clear to me, more crisply than it ever would be again, that summer was—no, not that summer was but that *life was.* The heavens are silky this evening, the swifts are rejoicing, the flowers are stretching out to me, I am walking toward them.

As I climb in the direction of the cemetery with my dog, I see a gigantic butterfly lying on the street. The body with its long wings is tipped onto one side. At first, I mistake it for an injured bird. Then I think that it is a giant hawk moth. I step closer and turn the animal over. It is the biggest butterfly I have ever seen in my life. As I lift it up, it begins to weakly flap its wings, which are already frayed on the edges.

I see the enormous, featherlike antennae on its head, the jagged orange links on its wings, the four mirroring eye spots, one on each wing section.

It is a great emperor moth, the biggest butterfly in Europe. Once, a few years earlier in Southern France, I saw its caterpillar. I carefully carry the moth in my hand to show my son, covering it with the other hand. Under my fingers, I feel a weak fluttering. The moth's wingspan is as big as my hand is long. I examine the feathered feelers with which the animal can perceive a single scent molecule from a lone female a dozen kilometers away and then follow this trace in order to mate. I am gripped by a feeling of melancholy.

My son asks: How rare is the butterfly? I say: So rare that you will perhaps never see it again in your lifetime. The caterpillars fatten themselves up all summer; the moths live for only a few weeks or even just a few days between April and June. How rare they have become. Did this one find his mate before he now must die? I set it down in the leaves of a laurel bush. Weakly he beats his wings, wearily he extends his furry legs under dull black round eyes in which all of these details sink, as though into a velvety and never-ending night. An ant arrives and touches the body that still pulses but is already near to being a corpse. The animal is a longtime foreigner, a last aristocrat of oversized proportions that burned up his energy during nightlong fluttering about the lamp outside the house of the mayor's widow. All of these encounters with scattered beings, all of these happy and melancholy meetings, each of which could be the last, allow me to encounter an aspect of myself—and a characteristic of the world.

The emperor moth is a body, immensely large. When I encounter it and feel everything that this encounter entails, when I, as a fragile part of the "flesh of the world," greet, recognize, and try to offer a little protection to another part of that same world, all of this ceases to be a merely external event. It is a drama of aliveness itself, one that has no place, is not outside or inside, in the "mind" or in the "body." If I look properly, I do see not the outsides of beings, but their aliveness and how it is organized around them. This is seeing with the heart, which the poet Antoine de Saint-Exupery spoke about. The heart moves within its own space, wherein it is not important whether something appears as a body or as a thought. It is the poetic space from which all of reality derives its power, a space beyond all division, between the expression of a body and that which exists as an existential gesture in space. The flesh of the world—it is simultaneously body and its feelings. It is a breathing skirmish of poetic relations.

Almost no one has understood the extent to which this inner space of aliveness among all beings is produced by the butterflies and their fragile majesty other than the Danish poet Inger Christensen. The slow disappearance of butterflies from the dead, intensively farmed fields of the food industry is also a farewell to poetry. We have imposed upon ourselves an exile from the poetic realm. This is why we find it increasingly difficult to understand what is really meant by life and death, pain and happiness. In her small but significant volume, *Butterfly Valley— A Requiem*, Christensen writes:

> *Is this flickering of wings only a shoal*
> *of light particles, a quirk of perception?*
> *Is it the dreamed summer hour of my childhood*
> *shattered as by lightning lost in time?*
>
> *No, this is the angel of light, who can paint*
> *himself as dark mnemosyne Apollo . . .*[2]

And a little lower, she writes:

> *[. . .] I play*
> *the fritillary caterpillar, gather*
> *all the world's life forms into one.*
>
> *Then I can answer death when it arrives:*
> *I'm playing the brown wood nymph; dare I hope*
> *that I'm an image of eternal summer?*[3]

We are always already inside

Every body is an existential drama turned into flesh and blood. It is a triumph over the forces of decay that constantly tug at it, a temporary victory of the erotic striving for unity and fullness over the ponderousness of matter. Every body is thus not merely a physique but always also a visible psyche: The significance of everything that has happened to it emanates outward from it. (The Greek word *psyche* is the term for "soul." The Greek goddess Psyche—the wife of Eros—was regularly depicted with two butterfly wings on her back. "Psyche" is also a biological name for a species of butterflies and a genus of moths—and for an entomological journal. Imagine a tree that grows arrow-straight into the air from a steep, rocky hillside: Its growth embodies excessive strains on life's forces and it embodies the overcoming of those strains. Everything that can be described as a biochemical process has this inside, which is visible when one wants to see it. We are surrounded by gestures of life and also continuously bring them into being ourselves.

Philosophically speaking, one could say: What researchers have been calling cognition for a long time—thinking and perception within the natural world—is, in fact, poetic expression. And this expression is no idyll. It includes everything. In the natural world, aliveness appears as the poetic principle, and this includes birth and death, growth and decay, ecstasy and grief. The natural world is a home, not a site of salvation, though many people still make it into that nowadays. I can understand this nostalgia. But what draws us to the natural world is the fact that it encompasses the whole of aliveness, all of the jubilation and torment—the fact that it is the embodied side of the yearning to be and to come into form, the poetic space from which everything comes and to which everything returns.

We can recognize the manifestations of this pure aliveness everywhere. To do so, we must adjust our gaze and use what Mary Catherine Bateson, daughter of Gregory Bateson, once called "peripheral vision," in a metaphor drawn from our visual capacities. In order to distinguish objects in dim light, we look not with the main focus of our eyesight but with that part of our eye that can take in the greatest amount of brightness, even though it does not see things most clearly. This part is a small depression in the retina, thickly covered by visual pigment—the fovea. The fovea is not the point of sharpest focus, but the spot that is most sensitive. We use peripheral vision, foveal seeing, involuntarily in very weak lighting, in order to distinguish objects from one another.

We sense more with this kind of seeing than we can recognize clearly. In peripheral vision, we exchange greater clarity for a maximum aperture. We no longer see everything in focus before us, but we begin to see with the whole body, in a way—with the same sensibility that we use to perceive ourselves. Peripheral seeing means perceiving by turning away from the light and no longer taking in its brightness with our eyes, a bit like feeling it on the back of the neck instead. All other beings enter into this felt aura, but not as biological species—as brothers and sisters in a space of aliveness.

Space and body become transparent when we look with peripheral vision. When we see with our whole body, which in turn is perceived by the bodies of the others in the vibrating field of flesh. When we see with all of our senses, with our aliveness itself. In this way we can perceive the gestures of life. In the end, these are the parts of the world that we can grasp most clearly, because they are us.

In our cores we perceive nothing less than that: pure aliveness, unspoken, beyond words, aliveness from within, which can respond to the traces of aliveness among other things we encounter. This is what heals us when we are despairing, what streams toward us as power from the natural world. Life heals life. Nowadays, some troubled patients are treated with "animal-assisted therapy": An assistant places an animal on their bed, maybe a young chick, or a gentle but boisterous puppy. The patients' medicine consists of nothing more than a high dose of pure aliveness, undiluted.

The natural world is not a mirror of our soul or emotions. It is an embodied expression of the soul and of emotionality, an expression that reveals that the inside of reality consists of this feeling dimension. The natural world is the space in which everything is simultaneously corporeal and full of meaning, simultaneously composed of flesh and spirit. The delicate branches of the nut tree whose veinlike quality recall the fine black texture on the surface of a fresh walnut; the cherry tree whose glowing fall colors dream again the hues of the summer sun that sated it, thereby expressing the autumn as a memory of the summer—not as consciousness but as being: These are all ways in which aliveness appears in a pure form. It can only appear in a body, but if we look at it properly, this body becomes something incorporeal, an existential gesture.

At this point, we can go even further than what we said in chapter 5. The natural world is not a copy or reflection of our soul or emotions. It is also more than a spatially "externalized" psyche that we can walk through (as the eco-philosopher Paul Shepard has suggested). It is rather the space of pure aliveness. The natural world, the sensory fabric of countless bodies touching and interpenetrating one another, is the place where we can track the conversion of matter into imagination that lies at the root of all aliveness.

Poetry also works with pure aliveness. It is the shaping of aliveness within the medium of words, colors, sounds, volumes. Its purpose is not to describe or to analyze but to re-enliven. As such, music (for example) is not the depiction of feelings through audible symbols, but a distinct form of appearance for the principles of life, for its crescendos and decrescendos, for intensification, arpeggiation, isolation, consolidation. The space of aliveness can be entered and felt on the skin. We draw near to it through our senses. And just as an erotic experience is only made possible by the touch of skin, artistic experiences can only be realized through the senses. Here, too, aliveness reveals itself only in the form of a body. I comprehend it via sensory channels that are consequently more than just organs of perception. They are organs of meaning.

If we draw upon peripheral seeing, we unfold a sensorium for the poetic space of aliveness. We have an intuition for the potential that urges us toward life. And we can recognize that there is something like an absolute spectrum of forms for the gestures of life that can be expressed in many different ways. Synesthetes know this: For them, colors do not just have a hue, they have a texture, a sound, and a taste—they can, for example, be velvety, smooth, concave, or stale.

What all of these aspects have in common is their meaning as gestures of life. The philosopher Hermann Schmitz once observed that "Mozart, the color yellow and a skipping gait" all have the same existential value: that of being joyous and light. We know what it means to be joyous and light from our bodily experiences. We can share this with other bodies, with all bodies in the flesh of the world. We can only experience it as a body, even though it is an experience that exceeds the body. For this reason, poetry is as necessary as the air we breathe: Knowledge of aliveness cannot be conveyed by facts, but only by being infected with life. Lewis Hyde writes: "Some knowledge cannot survive abstraction, and to preserve this knowledge we must have art. The liquid light, the *nous*, the fecundity of nature, the feeling of the soul in ascent—only the imagination can articulate our apprehension of these things, and the imagination speaks to us in images."[4]

The feeling of the soul in ascent is the feeling that the desire for aliveness that fills the cosmos to the point of overflowing is being realized. This absolute quality of enlivened reality belongs to all of us, because we all—all beings with bodies—can share in the existential experience of levity expressed in the small child's frolicking, in the fawn's bounding, in the mosquitoes' buzzing in the balmy evening air, in the swifts' movements as they circle ever higher above the tower of the old castle. We all have access to the poetic space, the only space that encompasses all other spaces. Within it, we can invoke anew the experience of existence by creative means: with swiftly sketched lines, with nimble-footed verses, with a rising cadence, with a camera shot pointed at a street climbing high into the springtime evening, the veined web of a leaf reaching for the outside. We cannot grasp this aliveness and hold it tight. We can only witness it by passing it on.

Desire that remains desire

The poetic space is that which has not yet come to be, and that which never was allowed to be. It is an act of looking at things that urges them to transform, the destructive disturbance that is never satisfied, the eve of spring in the big city park when the blackbird first sings awkwardly from a leafless tree. For René Char, the French poet and close friend of Albert Camus, this was the very definition of the poetic: "The poem is love realizing desire that remains desire,"[5] he writes.

Everything that is alive dwells in the body of this desire. Desire alone allows the body to hold itself together within the vortex of passing and fading matter. This life wish is what composes a being, more than its materiality. Living is the desire for wholeness that remains desire and forms itself accordingly. It is the poetry of the body, from the tiniest cell on up: a gesture of willful existence in entirely my own way, in the face of an unalterable end shared by all.

But this existence is not defiance, not insurgency; it is the conversion of nonexistence into affirmation, the transformation of death into a desire of flesh and blood, a breathing figure with a unique nature that can be spoken to and caressed. The body—both the individual body and the great body of the natural world—is this desire. It is aliveness from the inside of aliveness that has not yet found its form, that the observers see as seeking a form in which to appear, insofar as they believe they understand it, insofar as they sense an impulse of aliveness. It is that which will never be perfected, because it exists but never finds a beginning —it simply seeks and desires.

The poetic space itself is already there before any bodies express it. It is present as a wish and a lack—just as the most intense experience of spring is, in truth, the awareness of a lack of something that is about to come. The branches and the green grass become transparent and flimsy in order to allow for the anticipation of some potential behind this lack. The biophysicist Stuart Kauffman calls this timid display the "adjacent possible."[6] In this way, he tries to describe the biosphere's enduring capacity for self-organization: During the evolution of an ecosystem, the adjacent possible is constantly being explored and expanded. The more alive an ecosystem is, the more this adjacent possible swells invisibly within it.

The fact that the adjacent possible is invisible is, of course, only true for the experience of the scientists taking measurements who refuse to appreciate their own perception with peripheral vision, thereby refusing to perceive the part of the world that is enlivened, a living and breathing other. Those who see themselves as part of the world perceive aliveness's desire for them in the form of a pleasant anticipation. They allow themselves to be infected with life.

The sensibility that one gains thereby is ultimately nothing less than the creative power of a part of the natural world that is capable of bringing forth new physical and material life-forms. A new species or a new niche emerges from the "imagination of an ecosystem." This imaginative power condenses into joy in our perceptions. The diverse ecosystem will let us discover things that we could never have imagined; it will show us feelings that we had only intuited before.

The poetic space exists as a need to come into being and to be seen, as the unfelt's demand to be felt, the invisible's demand to be brought into the light, the immaterial's demand to become tangible as skin feeling pain and timid joy. The poetic space is that which all imperfect beings long to fill and which they will never fill completely. Every creative act is an attempt to satisfy a lack by filling the poetic space with concrete form, with lived experience, with the gesture of a pulsing body; and yet thereby, every creative act only carries forth this desire for more life, for fulfillment, for healing.

The poetic space is the space of potential that waits to be given away. The poetic space is the space of the Not-Yet, the space of lack in which that which can come into being appears as desire, as the pain of absence, the denial of which leads to creation. It is the space that holds the trembling possibility that every being must accept in order to continue being itself and thereby to become other than it is. The poetic space is opened by every truly living gesture—the gesture of a creature or ecosystem or the gesture of a work of art that deals with the meaning of life—because this gesture calls upon us to be enlivened and to put our own aliveness at risk.

And so we come to a definition of the poetic, here in these final pages.

The poetic is what calls us to be enlivened. It is a call—a call that creates the irresistible wish to bestow life on others. We can perceive this call in an image like Piero della Francesca's *Maria Magdalena* in Arezzo, in which the beholder sees how Maria, the depicted figure, lost in bliss, glimpses something invisible and is herself seen in return by this invisible thing. The poetic is a young puppy who tosses a nut into the air with its snout and then catches it. It is a child whose parents greet it with kindness that speaks from their eyes; a melody that won't leave your head and makes your body dance; a poem that shows the world in a new light, which can only be received if we satisfy the desire to write poetry of our own.

In the poetic space, contradictions are not resolved as they are in real-world occurrences, where every body represents a successful negotiation of opposites in every moment. In the poetic space, the contradictions are simultaneously present, as in the whole of reality before it has been delineated into individuals—as though in a simultaneous overview of all individuals, all life gestures.

The poetic of the living space is thus deeply permeated with paradoxes. It is present only in absence. It is an immaterial gesture that is nevertheless bound to bodies. It is already there but only as potential. I wrote about this "whole in the smallest fragment" in my book *The Biology of Wonder*. The poetic is the gesture of Eros, which wishes to merge and to fuse and must therefore summon separation. It is and it is not. All moments and all existing things in one tiny moment of exquisite aliveness: One could say that eternal life is actually the experience of one's own completely unguarded aliveness in a moment.

Many classical definitions of the beautiful involve this coupling of incompatible things. The romantic philosopher Schelling speaks about the total overlap of the general and the individual—a way of experiencing the "infinite in the finite." The poet Friedrich Schiller calls it "perfection in freedom." All of these definitions are paradoxes because they define a whole that abstracts itself into the creative tension between individuals, open to future possibilities.

The important part is hidden in that which is unattainable—in that which is desire. Both aspects must be real, both must be true in order for them to contain the real world. The plants thrust upward anew every spring from their hard husks. The death that the squirrel dies during a hard winter is real. Both are equally valid. Both are necessary, and yet they exist in a painful tension. The ecosphere is combat *and* cooperation, it is body *and* spirit, it is machine *and* imagination—and it must be all of these things at once. This coexistence forces us to create anew through unheard-of imagination. Poetic action means letting yourself be inspired to resonance by pure aliveness—not so that you can stockpile pleasing goods, but so that you can produce aliveness yourself. Poetic action does not reply to death with control or suppression but with readiness, with the capacity for and the affirmation of constant birth. Poetry is our greatest means of transformation. Poetry is our most powerful instrument of love.

Poetic materialism

One could perhaps use the designation *poetic materialism* to describe the attitude of erotic ecology. The biosphere is material, and this matter acts according to those most general life principles that we call natural laws. And the world is meaningful—or spiritual—insofar as everything that happens influences the inner experience of aliveness, and insofar as everything encountered reveals itself as a living gesture and thereby brings an absolute value into the world. And both are only possible together. Here, too, we find a tension between two opposites typically thought to be incompatible. *Poetic materialism* means that meaning remains embedded within the body and that we cannot extract it without damaging this body. We can only comprehend it if we ourselves are enlivened, and consequently we cannot help but transform it, because we enter into new connections with other bodies.

The poetic space—the living space of a creative reality—does not lie outside of the world. Nor is it really somewhere in the world, the way that something might be inside a box. It is the whole world, but not its surface; it is what we can understand through this surface. And so nothing is inconsequential. Every surface counts. This is not a form of aestheticism; it is simply the recognition that every one of our attitudes has a direct effect on our living state. Christopher Alexander, an architect and artist fighting in the most radical way to have us recognize as beautiful those things that enliven us, is therefore fighting constantly for life-producing environments. For acts of destruction in the material realm are always also catastrophes in the poetic realm, because the real is in fact also the poetic, but with specific regard to its significance for our aliveness.

I thought about this in the summer of 2010, as I sat on the tetrapods of the pier in Sestri Levante. The ocean lay stretched out before me to the horizon, occupied on the right-hand side by the thin blue streak of the Ligurian coast as it bent toward French Provence in the west. The ocean so large, so blue. I thought of that damaged oil well in the Gulf of Mexico that summer, which was spilling its contents ceaselessly into the ocean and wouldn't be closed again until autumn—and even then, only possibly.

The swaying, rolling surges shuddered one after the next before bursting on the blunt concrete arms of the breakwaters and unraveling their masses in white foam. What kind of feeling would fill us, I asked myself, if the ocean were an endless, grimy pool, a gigantic reservoir of refuse that disgorged itself in the majestic rise and fall of a gray, frothy groundswell onto our shores, onto our densely populated coastlines lined with pleasure palaces? Such an ocean would be a reflection of our despair, and yet it would also still be a reflection of those primal moments of power, tenacity, patience, and transgression that move all things. Even sewage-slung shores would be beautiful and sublime; even they would immoderately overcome our pettiness toward things.

On a bright afternoon shortly after this moment at the shore, I found a few slender gladiolus at the edge of my path in the Ligurian hinterland. They were wildflowers that only grew here on this Apennine peninsula —*Gladiolus italicus*. From that point on, I always walked the path past them, for the rapturous celebration of a few brief days. They, too—a gift from the nothingness. I had discovered them, had seen them unfold, marveled at them, and saw them disappear again. Upon greeting them, and upon saying farewell, I sketched them. Not to suspend the passage of time—the fleeting moment, already dying in the moment of its birth; not to immortalize them, but as a smiling greeting of the encounter and of the parting, I sketched them. The never fully representable shadows, billows, and variations in hue on a single petal—the "petali," as the daughter of my friend, the school caretaker, said in the beautiful melody of her Italian. The longer I tried to observe these details, the more exact I was with them, the deeper I plunged into their infinitude, as though I were tracing a coastline one grain of sand at a time. Time and space began to fall silent.

There would probably be a correlation between the depth of detail in one pink sepal, one light-infused petal of the *Gladiolus italicus,* and the endless length and dissolution of the coastline, if I could trace it grain by grain. I did not merely hit upon a metaphor to describe the elimination of space, of distance, of separation—this moment of sublation actually took place. But I stayed by my few gladiolus for as long as they were blooming. I drew pores and specks of color, I strolled around bays, cliffs, stones, pebbles, grains, crystal lattice, molecules, electrons: And here at the very latest, at this boundary of the quantum dimension, distance was obliterated in the nonlocality of all positions.

It is impossible to measure the position of an elementary particle without simultaneously connecting it with your measuring instrument, and thereby with the entire universe. In the same way, the light grid of a petal's surface dissolves into the afterglow of a whole day. It was there, and it was in me, and I was in it, just as the charge and momentum of the electron (mathematically and experimentally verifiable) is related to the opposing spin of its brother on the other side of the universe.

THE VOICE OF HAPPINESS

Laughing saves us. Seeing the other side of things, their surreal and amusing side, or at least imagining it, helps us not to be torn apart, brushed away like dry leaves, to resist in order to make it through the night, even though it seems endless.

ROBERTO BENIGNI[1]

*I*n the summer of 2013, I often rode out to the Teufelssee. The lake was not far, less than fifteen minutes by bike along the long street through the pines, birches, and beeches of the Grunewald forest in Berlin.

I pedaled through the forest past the dark green, blue green, sunlight-illuminated trees, unmoving beneath the sky that for weeks had been a sanguine blue. The small body of water lay black and smooth in a treeless valley of the Grunewald, at the bottom edge of an expansive meadow, rimmed by alders and blooming brush. I always rode there when I couldn't write any more, when my stuffy apartment on the big streets and the clamor that came in through the open door of the balcony became overwhelming.

I never lingered long, rode purposefully to the lake, stripped there on the bank, stuffed everything into the side bag of my bike and slipped into the water. Sometimes I met a friend and we glided together below the lake's cold and supple skin.

We swam in one of the little sandy inlets that opened within the shrubbery on the shore, leading into a pool of cool greenish water surrounded by gypsywort, viper's bugloss, forget-me-nots, and young willowherb. We gave our weight to the water and, after a moment of brief hesitation, it carried us. Most of the time, we had to gasp a little in pleasant surprise, which was always new and also always the same, eagerly awaited and also a completely new beginning, and then we had to laugh. The water carried us with smooth hands, and we laughed loudly, for no reason. We laughed in pure joy.

Sometimes, when I would ride there in the evening and go into the lake again, it seemed to me as though the water were still completely full of the last time we touched. I would let myself be embraced and caressed by the water—still warm in the uppermost layer and slightly cooler below. Tiny seed husks crowded on the sheer surface; slow, inky-blue clouds swam in varying combinations across the broad heavens; a fish would sometimes jump from the depths, tracing a rapid arc above the flat face of the water before disappearing again below. One evening, I discovered that on the west side, below fallen willow trees, the water lilies had bloomed—not the smaller yellow lilies, but true, white water lilies with succulent centers of sticky-furry stamen surrounded by halos of creamy white petals.

The lake was completely filled with itself, and on its surface the lilies drifted, just like me. The lake touched me with its cool, tender skin, whose taut surface hesitantly shivered as I moved through it, swimming. The lake let me truly feel my own skin with its gentle pressure; its cool, then warm enveloping of me. It was full of a You that embraced, bathed, and touched, and this touch made me into myself, allowed me to feel every square inch of my body's surface, and I had to laugh again, laugh with happiness in the middle of the lake, alone and carried, and I turned onto my back and gazed into the heavens, held by a smooth, slightly swaying mirror. That was the high point of the summer.

I remember that once, as I was swimming back, near to the shore, in a spot where I could stand again, I encountered another swimmer, a kind-looking Japanese woman who was wearing an oversized pair of goggles for some reason. She was standing in the water with her

squarish, overlarge mask and staring agog at four young ducks pad-
dling toward her from a carpet of water lilies. And then her whole
face broke out in laughter; she stood there alone and could not stop
laughing. And with my wet skin and limbs surrounded by cool water,
I understood that laughter is an organ of happiness. Not of "humor."
With our laughter, we greet happiness, just as this Japanese woman
greeted the ducks and the You in their black button eyes, just as I
greeted the water and the You in this water. All that mattered was
laughter, was laughing in happiness.

We laughed. And the way we laughed was the way you laugh when
you greet a loved one, the way you laugh to show another person that
happiness exists without saying a word—the very same way you laugh
when you hold out a bowl overflowing with food to a hungry dog,
chuckling quietly at its state of absolute rapture: "Look—this is really
here, let's be happy together about all these gifts." This is the truth: The
laughter of a baby, cooing with joy rather than laughing at a good joke, is
the zero line of happiness. This line is the enduring birth horizon of the
world's childhood.

We can confidently revise the famous essays by all of the serious
philosophers, the conceptual thinkers like Immanuel Kant or Max
Scheler, who classify laughter as a reaction to "humor." For starters, we
can just shove them into the back of the drawer. These people thought
primarily about mental refinements. They didn't think about the fact
that it is the body that laughs, that exults in the moments of contact
when it grasps that other bodies really exist, seized by that joyous gasp
that, if we are honest, always accompanies the encounter with the most
recent glaze that has been spread over the oldest things. They haven't
swum in the Teufelssee in the evening when the water lilies are opening and the curious young ducks go looking for people and the water
caresses your skin with its shy fingers, so happy to be allowed to touch
something living.

It is the other way around. Laughing is not the voice of humor.
Laughing at a joke is just a variety of happiness. It is the happiness that
finds unconnectable oppositions brought together for a brief moment
on a summer evening: my body and the water, the shy animals and the

bashful bathers, my body and the lake that form a single contact surface together. Laughter is joy in the fact that there will always be a way, that it is beautiful to be real. Laughter arises when a spark of life has been recognized, saved, kindled, and passed on.

Laughter arises from the joy of experiencing that beauty has its own power to exist. That it asserts itself. That it sustains. That it greets me. That it recognizes me. That I recognize it also. That I am able to see into its eyes, into the eyes of the water lilies with their alluring center of pistils and stamen, of innocent male and dreamy female sex organs, into the curious eyes of the ducks, into the calm eye of the lake itself, which welcomingly opens its watery mirror to the hesitant heavens. Happiness is a tangible essence in space. It is cool and green and transparent and it caresses and carries me, so long as I don't do anything to prevent it. I just have to go on breathing. I don't have to do anything.

And I also understood that love is an act of truly encountering an other, a willing engagement of the self whose first action is to risk its own skin. Diving into the lake between wild mint and sorrel and feeling the water. Taking the lake gently in your hands and letting it run between your fingers. And this practice of love means always accepting and even desiring what is, means being healed by being accepted, and this acceptance is, of course, also painful, and yet this pain is so minor when compared with the birth that accompanies it, or rather: The pain is perhaps immense and unending, but it is offset by the fact that its other is birth—birth that comes from being desired—and it is good, even if it is only the tiniest imaginable birth to balance out all of the world's pain, because that which is born is desired and wished for and loved by eyes that say: Be.

Yes, this feels right. Coherent. In thought, in the throat, in the stomach, in the pores of the skin. It feels right, now and for always. And it is good this way. Even if it hurts. Even when the doors open, close—irreversible. After that—the unknown. Death. The eternal beginning. The "always in never." A summer evening. A spark of the endless summer, of the world's summer essence that is always there, in your gaze, in my laugh.

Oh yes, and this must then be how God tries to do it—we have all been present to it at least once. Now, at the culmination of this evening hour, in the complete nakedness of the world, it is so easy to understand. And yet it takes so long before this effortlessness is finally remembered. And allowed.

It is no different. Never more planned, more protected, more professional, more hardened. Never less happy. It is just a step, a pause, a turn: There it is, within me, the happiness that does not belong to me alone, but to the world. I was born with it, I recognize it again, it was there at the beginning, the cells' feeling of amazement that they should possess so much beauty that they would multiply and combine into something ever greater. The newest amazement at the world's oldest matter-of-course: to be able to become greater and more beautiful and to feel the desire to share greatness and beauty with open hands, to be present to the invisible spark that is the center and the whole, to be the spark itself.

I was born with this—you, too. All of us. It is, in fact, the energy with which every birth is conceived. It is the thing that pushes us into life, the deep inside that has no form and no name and that yearns and also confidently knows that this yearning is insatiable and will therefore always be revived anew, no matter how devastating death and breakdown was. And is. And will be.

This is the moment that floods our skin when it sinks into the pillowy green weight of the Teufelssee, the moment in which the pointed ovals of the water lily's petals emerge in the soft perceptions of our body— their whiteness so unblemished by the void, their yellow in its inmost nourishing to our perceptions, like a warm egg yolk.

This is the reason the woman with the big goggles laughs while the young ducks paddle around her, curious, amazed, bathed in trust. All of this is the encounter with a living other that offers life, with a You that makes us into ourselves, into something that is, in turn, a You for the other. A feeling countenance that gifts trust.

We ourselves are this living guarantee, there in our inmost core, which is completely empty and then fully of the next moment, always an Even So, always a Yes. A bidding, Come. A Thank You.

Voilà.

In our inmost core, there it is.

As Marshall Rosenberg once described it, it is the way children feel when they feed a hungry duck. A game that makes both parties happy because it consists of nothing other than the exchange of gifts.[2]

Naked. Unarmored. Curious. Courageous.

Now.

VARESE LIGURE AND WESTEND, BERLIN
JANUARY 2012–FEBRUARY 2014

— *thanks* —

This book has been put together from a variety of threads that I would neither have found nor been able to follow without my friends' help. The thoughts on these pages are a weave that belongs to all of us. May the best of it be given back to you all.

Particularly I thank David Bollier for his perseverance in getting my ideas out to an English-speaking public. I am deeply grateful to Margo Baldwin for following my work and providing a home for it at Chelsea Green. Rory Bradley translated my language with lyrical sensitivity. Working with him felt like a second birth. The Bogliasco Foundation, Bogliasco and New York, gracefully accommodated my "Biopoetics" research project and offered me living space and time to think and feel. The Mesa Refuge, Point Reyes, California, welcomed me to the heart of American Nature Writing. I thank Kalevi Kull for the poetic bios and for the unstoppable loyalty, Stuart Kauffman and Katherine Peil for crucial support and friendship; Jeremy Lent and Elizabeth Ferguson for their unconditional hospitality, including a view of the Bay; Hildegard Kurt for radical goodness; Annelie Keil for the hope that does not die; Rainer Hagencord for companionship in the world inscape; David Abram for the fireflies; Per Espen Stoknes and Per Ingvar Haukeland for paths through the Norwegian wilderness; Giovanni Gotelli and Marzia Bancomini for providing the house in Via Colombo; Claudia Marenco for the laughter; Luciano Marcello, because I once managed to pay for his espresso; and Silvia Badino for finding my place at the Ligurian Sea.

Most deeply I thank my children Emma and Max for constantly challenging my thoughts with their living existences. I owe much of what I have learned about a practice of love to them, and I hope I will be able to give the best of it back to them. Finally I thank Erbse, my poodle friend, for never saying no, and nevertheless determining my life.

— notes —

EPIGRAPHS

1. Antonio Gramsci, "Socialism and Culture," in *The Gramsci Reader: Selected Writings 1916–1935*, ed. David Forgacs (New York: New York University Press, 2000), 59.

FOREWORD

1. Dylan Thomas, *The Collected Poems of Dylan Thomas*, "The Force That through the Green Fuse" (London: Weidenfeld and Nicolson, 2014), 43.

PRELUDE: THE CARRYING CAPACITY OF AIR

1. Theodor Lessing, *Meine Tiere* (Berlin, Germany: Matthes & Seitz, 2004), 142. Translation by Rory Bradley.
2. *Rondone* and *rondine* can be traced back to *hirundo*, the Latin word for swallow. Etymologists suspect that it is related to the Sanskrit root "har" or "ghar," which mean "to take" or "to grasp." See "L'hirondelle et la literature," *Les hirondelles*, http://www.hirondelle.oiseaux.net/litterature.html
3. J. M. Coetzee, *The Lives of Animals*, The Tanner Lectures on Human Values, delivered at Princeton University, October 15–16, 1997, http://tannerlectures.utah.edu/_documents/a-to-z/c/Coetzee99.pdf, 131.

PART ONE: I

1. Cees Nooteboom, *Gesammelte Werke, Band 6. Auf Reisen 3: Afrika, Asien, Amerika, Australien* (Frankfurt am Main, Germany: Suhrkamp-Verlag, 2004), 102. Translation by Andreas Weber.

CHAPTER ONE: TOUCH

1. John Muir, *John of the Mountains: The Unpublished Journals of John Muir*, ed. Linnie Marsh Wolfe (Madison: University of Wisconsin Press, 1938), 92.
2. Natalie Knapp, *Kompass neues Denken: Wie wir uns in einer unübersichtlichen Welt orientieren können* (Reinbek bei Hamburg, Germany: Rowohlt, 2011), 126.
3. Ibid., 94.

4. Gaston Bachelard, *L'eau et les rêves: Essai sur l'imagination de la matière* (Paris: Librairie José Corti, 1989), 32. Translation by Andreas Weber and Rory Bradley.

5. Muir, *John of the Mountains*, 165.

6. Sergeji L. Rubinstein, *Sein und Bewusstsein: die Stellung des Psychischen im allgemeinen Zusammenhang der Erscheinungen in der materiellen Welt [Being and Consciousness: On the Place of the Psychic in the General Interconnection of the Phenomena of the Material World]* (Berlin: Akademie-Verlag, 1970). Translation by Rory Bradley.

7. Quoted in Llewellyn Vaughan-Lee, *Love Is a Fire: The Sufi's Mystical Journey Home* (Point Reyes Station, CA: Golden Sufi Center, 2000), 22.

8. Gregory Bateson, "Form, Substance, and Difference," in *Steps to an Ecology of Mind: Collected Essays in Anthropology, Psychiatry, Evolution, and Epistemology* (New York: Ballantine Books, 1972), 448–66.

9. "Girolamo Fracastoro," *Wikipedia*, last modified January 4, 2017, https://de .wikipedia.org/wiki/Girolamo_Fracastoro.

CHAPTER TWO: DESIRE

1. Loren Eiseley, "The Little Treasures," in *Another Kind of Autumn* (New York: Scribner, 1977), 56–57. Reprinted with the permission of Scribner/Galley, a division of Simon & Schuster, Inc. from *Another Kind of Autumn* by Loren Eiseley. (Illustrated by Walter Ferro.) Copyright © 1977 by Estate of Loren Eiseley, c/o Mabel Eiseley, Executrix of the Estate. Illustration copyright 1977 by Walter Ferro. All rights reserved.

2. Francisco J. Varela, "Organism—A Meshwork of Selfless Selves," in *Organism and the Origins of Self*, ed. Alfred I. Tauber (Dordrecht, Netherlands: Springer, 1991).

3. Joseph Freiherr von Eichendorff, from his poem "Mondnacht" ["Night of the Moon"]. Eichendorff is a very well-known poet of late German Romanticism, and this is one of his most famous lines.

4. Peter J. Turnbaugh, et al., "The Human Microbiome Project: Exploring the Microbial Part of Ourselves in a Changing World," *Nature* 449, no. 7164 (October 18, 2007): 804–10.

5. Y. Fujita, et al., "Studies of Nitrogen Balance in Male Highlanders in Papua New Guinea," *Journal of Nutrition* 116, no. 4 (April 1986): 536–44.

6. Colin Nickerson, "Of Microbes and Men," *Boston Globe*, February 25, 2008.

7. Ibid.

8. Ibid.

9. Lord Alfred Tennyson's poem, "In Memoriam."

10. This phrase comes from Charles Darwin (citing the Swiss botanist, Augustin Pyrame de Candolle) during his first presentation of the theory of evolution before the Linnean Society in 1858.

11. Alexandra Bot and José Benites, "The Importance of Soil Organic Matter: Key to Drought-Resistant Soil and Sustained Food and Production," *FAO Soils Bulletin* 80 (2005), 1.

12. Gerald Hüther, in discussion with the author, May 8, 2010.

13. For more on the Three Laws of Desire, see Andreas Weber, *The Biology of Wonder: Aliveness, Feeling, and the Metamorphosis of Science* (Gabriola Island, Canada: New Society Publishers, 2016), 28–31.

CHAPTER THREE: DEATH

1. Ernst Jünger and Gershom Scholem, "Briefwechsel 1975–1981," *Sinn und Form* 61, no. 3, 2009: 293–302.

2. W. H. Auden, "New Year Letter" (1940).

3. Richard Rohr, *The Naked Now: Learning to See as the Mystics See* (New York: Crossroad Publishing Company, 2009), 11.

4. Alan W. Watts, *The Two Hands of God: The Myths of Polarity* (New York: George Braziller, 1963), quoted in Rohr, *The Naked Now*, 143.

5. Rainer Maria Rilke, "Closing Piece."

6. Gerard Manley Hopkins, "As Kingfishers Catch Fire," in *The Poems of Gerard Manley Hopkins* (London: Oxford University Press, 1918).

7. Hans Jonas, *The Phenomenon of Life: Toward a Philosophical Biology* (Evanston, IL: Northwestern University Press, 2001).

8. Theodor Adorno, *Aesthetic Theory*, trans. Robert Hullot-Kentor (London: Continuum, 1997), 73.

9. Sam Keen, "Foreword," in Ernest Becker, *The Denial of Death* (New York: Free Press, 1973), xiii.

10. Ibid., xv.

11. "NVC Marshall Rosenberg—San Francisco Workshop—Full Length English Subtitles Transcription," from The Basics of Nonviolent Communication: A 1-Day Introductory Workshop, San Francisco, 2000, posted by Centrum Nadania, October 27, 2015, https://www.youtube.com/watch?v=l7TONauJGfc.

12. Octavio Paz, *The Double Flame: Love and Eroticism*, trans. Helen Lane (New York: Harcourt Brace & Company, 1995), 274.

13. Ibid.

14. Gary Snyder, *The Practice of the Wild: Essays* (New York: North Point Press, 1990), 21.

15. Kimberlee Roth and Freda B. Friedman, *Surviving a Borderline Parent: How to Heal Your Childhood Wounds and Build Trust, Boundaries, and Self-Esteem* (Oakland, CA: New Harbinger Publishers, 2003), 139.

16. Becker, *The Denial of Death*, 153.

17. Wilhelm Reich, *The Murder of Christ: The Emotional Plague of Mankind* (London: Souvenir Press, 1975), 39.

18. Becker, *The Denial of Death*, 173.

19. David Schnarch, *Passionate Marriage: Love, Sex, and Intimacy in Emotionally Committed Relationships* (New York: W. W. Norton, 1997), 300.

20. Snyder, *The Practice of the Wild*, 5.

PART TWO: YOU

1. Walt Whitman, "Song of Myself," in *Song of Myself and Other Poems* (Berkeley, CA: Counterpoint), lines 1338–1340, quoted in Lewis Hyde, *The Gift: Creativity and the Artist in the Modern World* (New York: Vintage, 2007), 21.

CHAPTER FOUR: TRANSFORMATION

1. Simone Weil, "Toute séparation est un lien," in *La pesanteur et la grâce* (Paris: Plon, 1988), 164. Translation by Rory Bradley.
2. Natan P. F. Kellermann, "Epigenetic Transmission of Holocaust Trauma: Can Nightmares Be Inherited?" *Israel Journal of Psychiatry Related Science* 50, no. 1 (2013): 33–39; Louise J. Kaplan, *No Voice Is Ever Wholly Lost* (New York: Simon & Schuster, 1995), 222.
3. Paz, *The Double Flame*, 2.
4. Francisco J. Varela, "Organism—A Meshwork of Selfless Selves," in *Organism and the Origins of Self*, ed. Alfred I. Tauber (Dordrecht, Netherlands: Springer, 1991).
5. Helmuth Plessner, *Die Stufen des Organischen und der Mensch* (Berlin and Leipzig: De Gruyter, 1928), 22, quoted in Thomas Ebke, *Lebendiges Wissen des Lebens: Zur Verschränkung von Plessners philosophischer Anthropologie und Canguilhems historischer Epistemologie* (Berlin: Akademie Verlag, 2012), 92. Translation by Rory Bradley.
6. Albert Schweitzer, *Civilization and Ethics: The Philosophy of Civilization Part II*, trans. John Naish (London: A. & C. Black, Ltd., 1923), 253. Translation modified by Rory Bradley.
7. Gregory Bateson and Mary Catherine Bateson, *Angels Fear: Towards an Epistemology of the Sacred* (New York: MacMillan, 1987), 26–30.
8. Pablo Neruda, "Oh, Earth, Wait for Me," *Isla Negra*, trans. Alistair Reid, translation copyright 1981 by Alastair Reid, reprinted by permission of Farrar, Straus and Giroux; original Spanish: Pablo Neruda, "Oh Tierra, esperáme", *Memorial De Isla Negra*, copyright 1964, Fundación Pablo Neruda.
9. Paz, *The Double Flame*, 3.
10. Tomas Tranströmer. "From March 1979," in *New Collected Poems*, trans. Robin Fulton (Northumberland, U.K.: Bloodaxe Books, 2011). Reproduced with permission of Bloodaxe Books.
11. Shierry Weber Nicholsen, *The Love of Nature and the End of the World: The Unspoken Dimensions of Environmental Concern* (Cambridge, MA: MIT Press, 2002), 30.
12. Gary Snyder, "Unnatural Writing" in *A Place in Space: Ethics, Aesthetics and Watersheds* (Washington, D.C.: Island Press, 1996), 168, quoted in Nicholsen, *The Love of Nature and the End of the World*, 30.
13. Johann Wolfgang von Goethe, *Maxims and Reflections*, trans. Elisabeth Stopp (New York: Penguin Books, 1998), 77.

14. Goethe, quoted in Nicholsen, *The Love of Nature and the End of the World*, 82.
15. Jacques Derrida, "Che cos'è la poesia?" in *A Derrida Reader: Between the Blinds*, trans. Peggy Kamuf (New York: Columbia University Press, 1991), 221–40.
16. Jacques Derrida, "Qu'est-ce que la poésie? / Che cos'è la poesia / Was ist Dichtung/ What is poetry?" trans. Maurizio Ferraris, Alexander Garcia Düttmann, and Peggy Kamuf (Berlin: Brinkmann & Bose, 1990).

CHAPTER FIVE: EMBRACE

1. Quoted in Fabrice Midal, *Et si de l'amour on ne savait rien?* (Paris: Albin Michel, 2010), 137. Translation by Andreas Weber and Rory Bradley.
2. On the concept of "surplus" and the imaginative dimension of all life processes, see chapter 4 and also Varela, "Organism: A Meshwork of Selfless Selves," 79–107.
3. See, for example, Daniel Stern, *The Interpersonal World of the Infant: A View from Psychoanalysis and Developmental Psychology* (New York: Basic Books, 1985).
4. Andrew N. Meltzoff and M. Keith Moore, "Explaining Facial Imitation: A Theoretical Model," *Early Development and Parenting* 6 (1997), 179–92.
5. Schnarch, *Passionate Marriage*, 186.
6. Nicholsen, *The Love of Nature and the End of the World*, 23.
7. Christopher Quarch, *Hin und weg! Verliebe dich ins Leben* (Göttingen, Germany: J. Kamphausen, 2009), 49. Translation by Rory Bradley.
8. Midal, *Et si de l'amour on ne savait rien?*, 120. Translation by Andreas Weber.
9. Rohr, *The Naked Now*, 160.
10. Ibid., 144.
11. Knapp, *Kompass neues Denken*, 154. Translation by Rory Bradley.
12. Midal, *Et si de l'amour on ne savait rien?*, 11. Translation by Rory Bradley.
13. Schnarch, *Passionate Marriage*, 300.
14. Becker, *The Denial of Death*, 160.
15. Ibid., 165.
16. Ibid., 270.
17. Albert Camus, *The Rebel: An Essay on Man in Revolt*, trans. Anthony Bower (New York: Vintage Books, 1960), 306.
18. Albert Camus, *Lyrical and Critical Essays*, trans. Ellen Conroy Kennedy (New York: Vintage, 1968), 104.
19. Nicholsen, *The Love of Nature and the End of the World*, 102.
20. Varela, "Organism—A Meshwork of Selfless Selves."
21. Paul Valéry, "Analaects," in *Collected Works of Paul Valéry*, Vol. 14, trans. Stuart Gilbert (Princeton: Princeton University Press, 2008), 263.
22. Marina Zwetajewa, *Neuf lettres avec une dixième retenue et une onzième reçue* (Sauve, France: Clémence Hiver, 1991), 53, in Midal, *Et si de l'amour on ne savait rien?*, 137: "Aimer, c'est voir un homme comme Dieu l'a conçu et comme ses parents ne l'ont pas réalisé." Translation by Rory Bradley.

23. George Berkeley, *A Treatise Concerning the Principles of Human Knowledge* (Sheba Blake Publishing, 2014), 72.
24. Paz, *The Double Flame*, 273–74.
25. Schnarch, *Passionate Marriage*, 55.
26. Georges Bataille, *Erotism: Death and Sensuality*, trans. Mary Dalwood (San Francisco, CA: City Lights Books, 1962), 18: "Stripping naked [*mise à nu*] is seen in civilizations where the act has full significance if not as a simulacrum of the act of killing [*mise à mort*], at least as an equivalent shorn of gravity."

CHAPTER SIX: A PLAY OF FREEDOM

1. Joseph Campbell, quoted by Rosenberg, in "The Basics of Nonviolent Communication."
2. Scott Turner, personal message. For an in-depth analysis of play among animals see Andreas Weber, *Mehr Matsch: Kinder brauchen Natur* (Berlin: Ullstein-Verlag, 2011).
3. Alice Miller, *The Drama of the Gifted Child: The Search for the True Self*, trans. Ruth Ward (New York: Basic Books, 2008), 52.
4. Ibid., 52.
5. Angela Ebert and Murray J. Dyck, "The Experience of Mental Death: The Core Feature of Complex Posttraumatic Stress Disorder," *Clinical Psychology Review* 24, no. 6 (October 2004): 617–35.
6. Susan Forward, *Toxic Parents: Overcoming Their Hurtful Legacy and Reclaiming Your Life* (New York: Bantam Books, 1989), 5–6.
7. "Mental Disorders Affect One in Four People," World Health Organization, 2001, http://www.who.int/whr/2001/media_centre/press_release/en.
8. Yueqin Huang et al., "DSM–IV Personality Disorders in the WHO World Mental Health Surveys," *The British Journal of Psychiatry* 195, no. 1 (July 2009): 46–53.
9. Annelie Keil, in conversation with the author, December 2012.
10. Marie-France Hirigoyen, *Stalking the Soul: Emotional Abuse and the Erosion of Identity* (New York: Helen Marx Books, 2000).
11. Forward, *Toxic Parents*, quoted in Paul T. Mason and Randi Kreger, *Stop Walking on Eggshells: Taking Your Life Back When Someone You Care about Has Borderline Personality Disorder*, 2nd ed. (Oakland, CA: New Harbinger Publishers, 2010), iii.
12. Miller, *The Drama of the Gifted Child*, 34–35.
13. Ibid., 104.
14. Annelie Keil, "Krankheit als unvollendete Schöpfungstat," in *Chaos—Schöpfung—Evolution: Was die Welt im Innersten zusammenhält*, ed. Dieter Jarzombek (Berlin: LIT, 2011), 71–93. Translation by Rory Bradley.
15. Virginia Satir, *The New Peoplemaking* (Palo Alto, CA: Science and Behavior Books, 1988).
16. David Abram, *The Spell of the Sensuous: Perception and Language in a More-Than-Human World* (New York: Vintage, 1996), 264.

17. Abraham Maslow, "The Need to Know and the Fear of Knowing," *Journal of General Psychology* 68, no. 1 (1963): 111–25, quoted in Becker, *The Denial of Death*, 52.

18. Leonard Cohen, "Anthem," *The Future* (New York: Columbia, 1992).

19. Becker, *The Denial of Death*, 198.

20. Ibid., 159

21. Ibid., 126.

22. Wilfred Bion, quoted in Nicholsen, *The Love of Nature and the End of the World*, 161.

23. Hildegard Kurt, *Wachsen! Über das Geistige in der Nachhaltigkeit* (Stuttgart: Mayer, 2010), 128. Translation by Rory Bradley.

24. Paulo Coelho, *Warrior of the Light: A Manual*, trans. Margaret Jull Costa (New York: HarperCollins, 2003).

25. Rainer Maria Rilke, "Requiem." Translation by Rory Bradley.

26. Reich, *The Murder of Christ*, 35.

27. Ibid., 39.

28. Alice Miller, *For Your Own Good: Hidden Cruelty in Child-Rearing and the Roots of Violence* (New York: Farrar, Straus & Giroux, 1990), 98.

PART THREE: WE

1. Cited in Nicholsen, *The Love of Nature and the End of the World*, 120.

CHAPTER SEVEN:
THE THOUGHT OF THE SOUTHERN MIDDAY

1. Camus, *The Rebel*, 274.

2. Michel Onfray, *L'ordre libertaire: La vie philosophique d'Albert Camus* (Paris: Flammarion, 2013), 541. Translation by Andreas Weber and Rory Bradley.

3. Camus, *The Rebel*, 258.

4. Ibid., 296.

5. Ibid., 273.

6. Ibid., 255.

7. Ibid., 297.

8. Eugenio Montale, *The Collected Poems of Eugenio Montale, 1925–1977*, trans. William Arrowsmith (New York: W. W. Norton & Co., 2012), 31–33.
    Original Italian: *Ossi di Seppia* by Eugenio Montale. 1984 © Arnoldo Mondadori Editore S.p.A, Milano. 2015 © Mondadori Libri S.p.A, Milano.
    English Translation: "Cuttlefish Bones." Copyright 1948, 1925 by Arnoldo Mondadori Editore S.p.A., Milano. Translation copyright (c) 1992 by the Estate of William Arrowsmith, from *Collected Poems of Eugenio Montale 1925–1977* by Eugenio Montale, edited by Rosanna Warren, translated by William Arrowsmith. Used by permission of W. W. Norton & Company, Inc.

9. Camus, *The Rebel*, 302–3.

10. Iris Radisch, *Camus: Das Ideal der Einfachheit; Eine Biographie* (Reinbek bei Hamburg, Germany: Rowohlt, 2013), 155. Translation by Rory Bradley.

11. Rohr, *The Naked Now*, 132.

12. Helmut Leitner, *Pattern Theory: Introduction and Perspectives on the Tracks of Christopher Alexander* (United States: Helmut Leitner, 2015).

13. Weber, *The Biology of Wonder*. See also Bruce H. Weber, James D. Smith, and David J. Depew, *Entropy, Information and Evolution: New Perspectives on Physical and Biological Evolution* (Cambridge, MA: MIT Press, 1988).

14. John Maynard Keynes, "Economic Possibilities for Our Grandchildren," in *Essays in Persuasion* (New York: Norton, 1963), 358–73.

15. Francisco J. Varela, *Ethical Know-How: Action, Wisdom, and Cognition* (Stanford, CA: Stanford University Press, 1999).

16. Francisco J. Varela, "Intimate Distances: Fragments for a Phenomenology of Organ Transplantation," *Journal of Consciousness Studies* 8, no. 5–7 (2001), 259–71.

17. The metaphor of the dark mountain goes back to the Dark Mountain Project, a loose literary collation that seeks to develop a perspective on the crisis of the natural world uninfluenced by utopias of salvation. See http://dark-mountain.net.

18. Francisco J. Varela, Evan T. Thompson, and Eleanor Rosch, *The Embodied Mind: Cognitive Science and Human Experience* (Cambridge, MA: MIT Press, 1991).

19. Robinson Jeffers, quoted in "Robinson Jeffers," *Wikipedia*, last updated January 2, 2017, http://en.wikipedia.org/wiki/Robinson_Jeffers.

CHAPTER EIGHT: SHARING

1. Rosenberg, "The Basics of Nonviolent Communication."

2. Hyde, *The Gift*, 66–67.

3. Ibid., 212, 195.

4. Ibid., 48.

5. Ibid., 13.

6. Andreas Weber, *Biokapital; Die Versöhnung von Natur, Ökonomie und Menschlichkeit* (Berlin: Berliner Taschenbuch Verlag, 2008); Andreas Weber, *Enlivenment: Towards a Fundamental Shift in the Concepts of Nature, Culture, and Politics* (Berlin: Heinrich-Böll-Stiftung, 2013).

7. Cormac McCarthy, *The Crossing* (New York: Vintage Books, 1995), 127.

8. Snyder, *The Practice of the Wild*, 20.

9. Becker, *The Denial of Death*, 173

10. Otto Rank, cited in Becker, *The Denial of Death*, 174.

11. Nicholsen, *The Love of Nature and the End of the World*, 87.

12. Ibid., 91.

13. Hans Jonas, "Der Gottesbegriff nach Auschwitz. Eine jüdische Stimme," in *Philosophische Untersuchungen und metaphysische Vermutungen* (Frankfurt am Main, Germany: Insel Verlag, 1992), 190–208. Translation by Rory Bradley.

14. Thomas Merton, *Conjectures of a Guilty Bystander* (New York: Doubleday, 2009), 155, quoted in Llewellyn Vaughan-Lee, *Prayer of the Heart in Christian and Sufi Mysticism* (Point Reyes Station, CA: Golden Sufi Center, 2012), xiv.

15. Hans Jonas, "Matter, Mind and Creation," in *Mortality and Morality: A Search for Good after Auschwitz* (Evanston, IL: Northwestern University Press, 1996), 191.

16. Simone Weil, quoted in Magdalena S. Gmehling, "Aus der Liebe leben, and der Liebe sterben: Verwandelt durch Gottesliebe: Heilige Theresa vom Kinde Jesu und Simone Weil," in *Einsicht: Römisch-Katholische Zeitschrift*, Jahrgang 28, Nr. 2, München 1984. Translation by Rory Bradley.

17. Rohr, *The Naked Now*, 103.

18. Hans Jonas, "Geist, Natur und Schöpfung," Kosmologischer Befund und kosmogonische Vermutung, in: *Hans-Peter Dürr and Walter Ch. Zimmerli*, eds., Geist und Natur, Über den Widerspruch zwischen naturwissenschaftlicher Erkenntnis und philosophischer Welterfahrung (München: Scherz, 1989) 61. Translation by Rory Bradley. Jonas refers to a term coined by the German philosopher Ludwig Klages in the 1920s.

CHAPTER NINE: THE HEAVENS, NOW

1. Jorge Luis Borges, "The Aleph," in *The Aleph: Including the Prose Fictions from The Maker*, trans. Andrew Hurley (New York: Penguin Classics, 2000), 130–31.

2. Inger Christensen, *Butterfly Valley: A Requiem*, trans. Susanna Nied (New York: New Directions, 2004), 3. Copyright ©2003, 2004 by Susanna Nied; Copyright © 2001 by Susanna Nied and Dedalus Press; Copyright © 1963, 1982 by Inger Christensen and Gyldendal; Copyright © 1989, 1991 by Inger Christensen and Broendums Forlag. Reprinted by permission of New Directions Publishing Corp.

3. Ibid., 15.

4. Hyde, *The Gift*, 287.

5. René Char, cited in Midal, *Et si de l'amour on ne savait rien?*, 7: "Le poème est l'amour réalisé du désir demeuré désir." Translation by Rory Bradley.

6. Stuart Kauffman, *Investigations* (New York: Oxford University Press, 2000), 142.

POSTLUDE: THE VOICE OF HAPPINESS

1. Roberto Benigni, "Benigni. Il mio film sdrammatico." *Corriere della Sera* (December 19, 1997): 31. Translation by Andreas Weber and Rory Bradley.

2. Rosenberg, "The Basics of Nonviolent Communication."

— *index* —

ERRATA

Page 44 – each which include... (not 'each introduced with')

Page 44 – add:

> Don't hurry - "be still" and "wait on the Lord"
> Read the text out loud to yourself
> Ask yourself: is this a command? a warning? a promise?

Page 45 – add: Write down thoughts or summarize main points in the Margin; Create a visual reminder by adding a simple drawing to illustrate an idea

Pages 227, 228, 249 and 251 – Hebrew **שָׁלַח** not חָלַשׁ

Page 233 – Line 7, fallible NOT the word infallible

Page 241 – Add the two essentials:

> **The Sending Return** (Report/Recall) – Cf. Luke 9:10, Luke 10:17-24, and Christ's Ascension
>
> **The Sending Celebration** – Cf. Luke 15

Page 243 – Add credit: By Rev. Dr. John Hirsch, modified by Rev. Dr. Steve Sohns and Rev. Ted Benson

Page 295 – Correct Rev. Hennings "bio" (last ¶): *Hennings and his wife, Val, are members of **Faith** Lutheran Church in **Georgetown**. They have four children and eight grandchildren: Kim (Eric) Otten, with children Nathaniel, Matthew, and Joanna of Omaha, Nebraska; Rev. Paul Hennings, **(and Ellie)** with children Theo and Toby of Cedar Rapids, Iowa; Rev. Luke (Lisa) Hennings, with children Logan, Caleb, and Grant of Mesa, Arizona; and Erika Hennings of New York, New York.*

Page 297-298 – add (1st ¶) to Dr. David Maier's "Bio:" *Rev. Dr. David Maier was brought up in a loving Christian home with a family legacy of service to God and His Church. He served congregations in Michigan and Illinois from 1982 until June 2009 when he was elected President of the Michigan District, LCMS...* (last ¶) *David and his wife, Pat, reside in northern Michigan. They have been blessed with four children - Leah (Kyle), Joel (Shae), James, and Hannah (Garrett), and 5 grandchildren so far - Beckett, Avery, Grayson, Luke, and Ella Bliss.*

Page 298 – add (last ¶) to Dr. Paul Maier's Bio: *Paul and his wife Joan have four daughters - Laura (Brett), Julie, Krista (Zach), and Katie (Brad), and nine grandchildren - Christine, Nicole, Anna, Drew, Rachel, Mark, Ben, Adam, and Luke.*